Sir John Lubbock

Ants, Bees and Wasps

A Record of Observations on the Habits of the Social Hymenoptera

Sir John Lubbock

Ants, Bees and Wasps
A Record of Observations on the Habits of the Social Hymenoptera

ISBN/EAN: 9783337144012

Printed in Europe, USA, Canada, Australia, Japan

Cover: Foto ©berggeist007 / pixelio.de

More available books at **www.hansebooks.com**

THE INTERNATIONAL SCIENTIFIC SERIES.

ANTS, BEES, AND WASPS.

A RECORD OF OBSERVATIONS

ON THE

HABITS OF THE SOCIAL HYMENOPTERA.

BY

Sir JOHN LUBBOCK, Bart., M.P., F.R.S., D.C.L., LL.D.,

REVISED EDITION.

NEW YORK:
D. APPLETON AND COMPANY.
1897.

PREFACE.

This volume contains the record of various experiments made with ants, bees, and wasps during the past ten years. Other occupations and many interruptions, political and professional, have prevented me from making them so full and complete as I had hoped. My parliamentary duties, in particular, have absorbed most of my time just at the season of year when these insects can be most profitably studied. I have, therefore, whenever it seemed necessary, carefully recorded the month during which the observations were made; for the instincts and behaviour of ants, bees, and wasps are by no means the same throughout the year. My object has been not so much to describe the usual habits of these insects as to test their mental condition and powers of sense.

Although the observations of Huber, Forel, McCook, and others are no doubt perfectly trustworthy, there are a number of scattered stories about ants which are quite unworthy of credence; and there is also a large class in which, although the facts may be correctly recorded, the inferences drawn from them are very questionable. I have endeavoured, therefore, by actual experiments which any one may, and I hope others will, repeat and verify, to throw some light on these interesting questions.

The principal point in which my mode of experimenting has differed from that of previous observers has been that I have carefully marked and watched particular insects; and secondly, that I have had nests under observation for long periods. No one before had ever kept an ants' nest for more than a few months. I have one now in my room which has been under constant observation ever since 1874, *i.e.* for more than seven years.[1]

---

[1] I may add that these ants are still (March 1882) alive and well. The queens at least are now eight years old, if not more.

I had intended to make my observations principally on bees; but I soon found that ants were more convenient for most experimental purposes, and I think they have also more power and flexibility of mind. They are certainly far calmer, and less excitable.

I do not attempt to give anything like a full life-history of ants, but I have reproduced the substance of two Royal Institution lectures, which may serve as an introduction to the subject. Many of the facts there recorded will doubtless be familiar to most of my readers, but without the knowledge of them the experiments described in the subsequent chapters would scarcely be intelligible.

I have given a few plates illustrating some of the species to which reference has been most frequently made; selecting Lithography (as I was anxious that the figures should be coloured), and having all the species of ants drawn to one scale, although I was thus obliged in some measure to sacrifice the sharpness of outline, and the more minute details. I am indebted to Mr.

BATES, Dr. GÜNTHER, Mr. KIRBY, and Mr. WATERHOUSE, for their kind assistance in the preparation of the plates.

As regards bees and wasps, I have confined myself for want of space to the simple record of my own observations.

I am fully conscious that experiments conducted as mine have been leave much to be desired, and are scarcely fair upon the ants. In their native haunts and under natural conditions, more especially in warmer climates, they may well be expected not only to manifest a more vivid life, but to develop higher powers.

I think, however, that my volume will at least show the great interest of the subject, and the numerous problems which still remain to be solved.

HIGH ELMS, DOWN, KENT:
October 18, 1881

CONTENTS.

CHAPTER I

INTRODUCTION.

Position of ants in the Animal Kingdom—Ants divided into three families—Number of species—Mode of observation—Nests—Mode of marking ants—Stages in life of ants—Egg, larva, pupa, imago—Length of life—Structure of ants—Head, thorax, abdomen, antennæ, eyes, ocelli, mouth parts, legs, wings, sting—Origin of the sting—Character of ants—Wars among ants—Modes of fighting—Queen ants—Workers—Different classes of workers—The honey ant—Soldiers—Origin of the soldiers—Division of labour—Habitations of ants—Communities of ants—Food—Enemies—Character—Industry—Games—Cleanliness 1

CHAPTER II.

FORMATION AND MAINTENANCE OF NESTS.

Foundation of new nests—Doubts on the subject—Views of Huber, Blanchard, Forel, St. Fargeau, Ebrard—Experiments with queens—Foundation of a nest of *Myrmica* by two queens—Adoption of queens—Fertility of workers—Eggs laid by fertile workers always produce males—Queens seldom produced in captivity—Origin of difference between queens and workers—Longevity of ants—Arrangement of chambers in a nest—Division of labour—The honey ant . 30

CHAPTER III.

ON THE RELATION OF ANTS TO PLANTS.

Flowers and insects—Ants not so important in relation to flowers as bees, but not without influence—Ants seldom promote cross-fertilisation, and hence injurious to flowers—Modes by which they are excluded — Belt — Kerner — Aquatic plants — Moats — *Dipsacus* — Slippery surfaces—Gentian, snowdrop, cyclamen — Concealment of honey—*Antirrhinum, Linaria, Campanula, Ranunculus, Lamium, Primula, Geranium,* &c.—Protection of honey by thickets of spines or hairs—Protection by viscid secretions—*Silene, Senecio, Linnæa, Polygonum,* &c.—Milky juice—*Lactuca*—Nectaries on leaves—Leaf-cutting ants—Ants as tree guards—Importance of ants in destroying other insects—Harvesting ants—Solomon—The Mischna—Meer Hassan Ali-Sykes—Moggridge—Agricultural ants—Lincecum—McCook 50

CHAPTER IV.

RELATIONS TO OTHER ANIMALS.

Hunting ants—The Driver ants—Ecitons—Insects mimicking ants—Enemies of ants—Parisites—Mites—*Phora*—Domestic animals of ants—*Aphides*—Eggs of *Aphides* kept through the winter by ants—Blind beetles—Pets—Progress among ants—Relations of ants to one another—*Stenamma—Solenopsis*—Slave-making ants — *Formica sanguinea* — *Polyergus* — Expeditions of *Polyergus*—*Polyergus* fed by the slaves—*Strongylognathus*—Degradation of *Strongylognathus*—*Anergates*—Explanation of the present state of *Strongylognathus* and *Anergates*—Progress among ants—Phases of life—Hunting, pastoral, and agricultural species .   .   . 63

CHAPTER V.

BEHAVIOUR TO RELATIONS.

Mr. Grote on 'Morality as a necessity of society'—Behaviour of ants to one another—Statements of previous writers:

CONTENTS.

Latreille, St. Fargeau, Forel—Difference of character among ants—Experiments—Isolated combats—Neglect of companions if in trouble—Experiments with insensible ants—Drowned ants—Buried ants—Contrast of behaviour to friends and strangers—Instances of kindness—A crippled ant—A dead queen—Behaviour to chloroformed friends—Behaviour to intoxicated friends . . . 93

CHAPTER VI.

RECOGNITION OF FRIENDS.

Number of ants in a community—They all recognise one another—All others are enemies—Recognition after separation—Strange ants never tolerated in a nest—Experiments—Behaviour to one another after a separation of more than a year—Recognition unmistakable—How are they recognised?—Some naturalists have suggested by scent, some by a pass-word—Experiments with intoxicated ants—With pupæ removed from the nest and subsequently returned—Separation of a nest into two halves, and recognition as friends by the ants in each half of young bred in the other half—Pupæ tended by ants from a different nest treated as friends in the nest from which they were taken, and as strangers if put into the nest of their nurses—Recognition neither personal nor by means of a pass-word 119

CHAPTER VII.

POWER OF COMMUNICATION.

Statements of previous writers: Kirby and Spence, Huber, Franklin, Dugardin, Forel—Habit of bringing friends to food—Exceptional cases—Experiments to determine whether ants are brought or directed to stores of food—Scent—Sight—Experiments with different quantities of food—Ants which returned empty-handed and brought friends to assist 153

CHAPTER VIII.

THE SENSES OF ANTS.

PAGE

SIGHT:—Difficulty of understanding how insects see—Number of eyes—Two theories—Views of Müller, Grenacher, Lowne, Claparède—Appreciation of colour—Sensitiveness to violet—Perception of ultra-violet rays. HEARING:—Antennæ regarded by many entomologists as organs of hearing—Opinions as to whether ants, bees, and wasps hear—General opinion that bees and wasps can hear—Huber and Forel doubt in the case of ants—Experiments with ants—Forel's observations—Colonel Long—Mr. Tait—Structure of anterior tibia. THE SENSE OF SMELL . . . 182

CHAPTER IX.

GENERAL INTELLIGENCE

Statements of previous writers—Economy of labour—Experiments as to ingenuity in overcoming obstacles and economising labour—Experiments with bridges, embankments, and moats—Earthworks—Ingenuity in building nests—Difficulty in finding their way—Experiments with movable objects—Sense of direction—Experiments with rotating disks—Experiments with rotating table—Influence of light 236

CHAPTER X.

BEES.

Difficulty experienced by bees in finding their way—Communication between bees—Bees do not by any means always summon one another when they have discovered a store of food—Bees in strange hives—Infatuation of bees—Want of affection—Behaviour to queen—Sentinels—The sense of hearing—The sense of colour—Experiments with coloured papers—Power of distinguishing colours—Preference for blue—Influence of bees on the colours of flowers—Blue flowers—Paucity of blue flowers—Blue flowers of comparatively recent origin 274

CHAPTER XI.

WASPS.

Communication among wasps—Like bees, they by no means invariably bring companions when they have discovered a store of food—Courage of wasps—*Polistes gallica*—A tame wasp—Power of distinguishing colours—Wasps less guided by colour than bees—Industry of wasps—A day's work—Directness of flight of wasps . . . 311

APPENDICES 323

LIST OF ILLUSTRATIONS

PLATE I.

Fig. 1. *Lasius niger* ☿.
 " 2.   "   *flavus* ☿.
 " 3. *Formica fusca* ☿.

Fig. 4. *Myrmica ruginodis* ☿.
 " 5. *Polyergus rufescens* ☿.
 " 6. *Formica sanguinea* ☿.

PLATE II.

Fig. 1. *Atta barbara* ☿ major.
 " 2.   "   "   ☿ minor.
 " 3. *Pheidole megacephala* ☿ major.

Fig. 4. *Pheidole megacephala* ☿ minor.
 " 5. *Formica rufa*.

PLATE III.

Fig. 1. *Œcodoma cephalotes* ☿ major.
 " 2. *Œcodoma cephalotes* ☿ minor.

Fig. 3. *Stenamma Westwoodii* ☿.
 " 4. *Solenopsis fugax* ☿.

PLATE IV.

Fig. 1. *Camponotus inflatus* ☿.
 " 2. *Tetramorium cæspitum* ☿.

Fig. 3. *Strongylognathus testaceus* ☿.
 " *Anergates atratulus* ♀.

PLATE V.

Fig. 1. *Lasius flavus* ♀.
 " 2.   "   "   ♂.
 " 3.   "   "   larva.
 " 4.   "   "   pupa.
Fig. 5. *Beckia albinos*.

 " 6. *Aphis*.
 " 7. *Platyarthrus Hoffmanseggii*.
 " *Claviger foveolatus*

LIST OF THE

PRINCIPAL BOOKS AND MEMOIRS

REFERRED TO.

ANDRÉ, E.	Desc. des Fourmis d'Europe. Rev. et Mag. de Zool., 1874.
BATES, H W.	The Naturalist on the Amazons.
BELT, T.	. The Naturalist in Nicaragua.
BERT, PAUL	. 'Les animaux voient-ils les mêmes rayons lumineux que nous?' Arch. de Physiol, 1869.
BLANCHARD, E.	. Metamorphoses of Insects. Trans. by Duncan.
BOISSIER DE SAUVAGES, l'Abbé.	L'origine du Miel. Journ. de Physique, vol. i.
BÜCHNER, L.	. Mind in Animals.
BUCKLEY, S. B.	. On Myrmica molefaciens. Proc. Acad. Nat. Sci. Philadelphia, 1860.
BURMEISTER, H.	. Manual of Entomology.
CURTIS, J	On the Genus Myrmica. Trans. Linn. Soc., 1854.
DARWIN, C.	. Origin of Species.
DELPINO, F.	. Sui rapporti delle Formiche colle Tettigometre.
DEWITZ, H.	. Ueber Bau und Entwickelung des Stachels der Ameisen. Zeits. f. Wiss. Zoologie, vol. xxviii.
DUJARDIN, F.	. Obs. sur les Abeilles. Ann. des Sci. Nat., 1852.
EDWARDS, H.	. Notes on the Honey-making Ants. Proc. California Acad., 1873.
ELDITT, H. L.	Die Ameisen-Colonien u. deren Mitbewohner
EMERY, C.	Saggio di un ordinamento naturale dei Mirmicidei.

EMERY, C.	Le Formiche ipogei. Ann. Mus. Civ. di St. Nat. di Genova.
FOREL, A.	Fourmis de la Suisse.
GÉLIEU, J. DE	Le Conservateur des Abeilles.
GOULD, Rev. W.	Account of English Ants.
GRÄBER, VITUS	Die Tympanalen-Sinnesapparate der Orthopteren.
GREDLER, V.	Der zoologische Garten.
GRIMM	Die Myrmecophilen. Stettin. Ent. Zeits., 1845.
HAGENS, Herr VON	Ueber Ameisengäste. Berlin. Ent. Zeits., 1865.
„ „	Ueber Ameisen mit gemischten Colonien. Berlin. Ent. Zeits., 1867.
HEER, O.	Die Hausameisen Madeiras. Zürich. Nat. Ges., 1852.
HUBER, P.	Natural History of Ants.
HUXLEY, T. H.	On the Reproduction of Aphis. Trans. Linn. Soc., xxii. 1859.
KERNER, Dr. A.	Flowers and their Unbidden Guests. Trans. by Ogle.
KIRBY AND SPENCE	Introd. to Entomology.
LANDOIS, Dr. H.	Thierstimmen; also Zeits. für Wiss. Zool. 1867.
LANGSTROTH, L. L.	Treatise on the Honey Bee.
LATREILE, P.	Hist. Nat. des Fourmis.
LESPÈS, C.	Sur les Mœurs de Lomechusa paradoxa. Ann. des Sci. Nat., 1863.
LINCECUM, GIDEON	On the Agricultural Ant of Texas. Linn. Journal, 1861.
LONG, Col. C. G.	Central Africa.
LUBBOCK, Sir J.	On the Anatomy of Ants. Trans. Linn. Soc., 1879.
„	Ova and Pseudova of Insects. Phil. Trans., 1858.
„	Obs. on Ants, Bees, and Wasps. Parts 1–9, Linn. Journ., 1874–81.
„	On Some Points in the Anatomy of Ants. Micros. Soc., 1877.
LUND, M	Lettres sur les Habitudes de quelques Fourmis du Brésil. Ann. des Sci. Nat., xxiii. 1831.

McCook, H. C. Note on Adoption of a Queen Ant. Proc. Acad. Nat. Sci. Philadelphia, 1879.
„ . On the Nat. His. of the Agricultural Ant of Texas.
„ The Honey Ant of Texas.
Märkel, F. . . Beit. zur Kenntniss der unter Ameisen lebenden Insecten. Germar's Zeit. Ent., 1841.
Mayr, Dr. G. L. . Europ. Formiciden.
„ . . Leben und Wirken der einh. Ameisen.
Meinert, F. . . Bidrag til de Danske Myrers Naturhistorie. Kiöbenhaven, Dansk. vid. Selsk., 1861.
Meyer, J . Ueber conoonlose Ameisenpuppen. Stettin Ent. Zeit., 1854.
Müller, P. W. J. . Beiträge zur Naturgeschichte der Gattung Claviger. Germar's Mag. de Zool., 1818.
Ormerod, E. L. . Natural History of Wasps.
Rambert, M. . . Mœurs des Fourmis.
Robert, E. . . Observations sur les Mœurs des Fourmis. Ann. des Sci. Nat, 1842.
Roger, J. . . Beit. zur Kennt. der Ameisenfauna der Mittelmeerländer. Berlin. Ent. Zeit., 1857.
St. Fargeau, Lepeletier. Hist. Nat. des Hyménoptères.
Saunders, Edward Brit. Heterogyna and Foss. Hymenoptera. Trans. Ent. Soc., 1880.
Savage, T. S. . On the Habits of Driver Ants. Trans. Ent. Soc., 1847.
Schenk, Professor . Beschr. Nassau. Ameisenarten. Stettin. Ent. Zeit., 1853.
Siebold, C. T. von. Ueber das Stimm. und Gehörorgan der Orthopteren. Weissmann's Arch., 1844.
Smith, F. Cat. of Brit. Foss. Hymenoptera.
„ . . Essay on British Formicidæ. Trans. Ent. Soc., N.S. vol. iii. p. 98.
Sykes, Col. . Account of Pheidole providens. Trans. Ent. Soc., 1836.
Wesmael, C. Sur une nouv. Espèce de Fourmi du Mexique. Bull. de l'Acad. de Sci. de Bruxelles, 1838.
Westwood, J. O. . Modern Classification of Insects.
„ . Obs. on Typhlopone. Ann. Mag. Nat. Hist., 1841.

ANTS, BEES, AND WASPS.

CHAPTER

INTRODUCTION.

The Anthropoid apes no doubt approach nearer to man in bodily structure than do any other animals; but when we consider the habits of Ants, their social organisation, their large communities, and elaborate habitations; their roadways, their possession of domestic animals, and even, in some cases, of slaves, it must be admitted that they have a fair claim to rank next to man in the scale of intelligence. They present, moreover, not only a most interesting but also a very extensive field of study. They are divided into three families: the Formicidæ, Poneridæ, and Myrmicidæ, comprising many genera and a large number of species. In this country we have rather more than thirty kinds; but ants become more numerous in species, as well as individuals, in warmer countries, and more than a thousand species are known. Even this large number is certainly far short of those actually in existence.[1]

I have kept in captivity about half of our British species of ants, as well as a considerable number of foreign forms, and for the last few years have generally had from thirty to forty communities under observation. After trying various plans, I found the most convenient method was to keep them in nests consisting of two plates of common window glass, about ten inches square, and at a distance apart of from $\frac{1}{10}$ to $\frac{1}{4}$ of an inch (in fact just sufficiently deep to allow the ants freedom of motion), with slips of wood round the edges, the intermediate space being filled up with fine earth. If the interval between the glass plates was too great, the ants were partly hidden by the earth, but when the distance between the plates of glass was properly regulated with reference to the size of the ants, they were open to close observation, and had no opportunity of concealing themselves. Ants, however, very much dislike light in their nests, probably because it makes them think themselves insecure, and I always therefore kept the nests covered over, except when under actual

---

[1] I have had some doubt whether I should append descriptions of the British species. On the whole, however, I have not thought it necessary to do so. They are well given in various entomological works: for instance, in Smith's *Catalogue of British Fossorial Hymenoptera*; Saunders' *Synopsis of British Heterogyna*; and in Mayr's *Die Europäischen Formiciden*, all of which are cheap and easily procurable. I have, however, given figures of the principal species with which I have worked.

MODE OF OBSERVATION.

observation. I found it convenient to have one side of the nest formed by a loose slip of wood, and at one corner I left a small door. These glass nests I either kept in shallow boxes with loose glass covers resting on baize, which admitted enough air, and yet was impervious to the ants; or on stands surrounded either by water, or by fur, with the hairs pointing downwards. Some of the nests I arranged on stands, as shown in

FIG. 1.

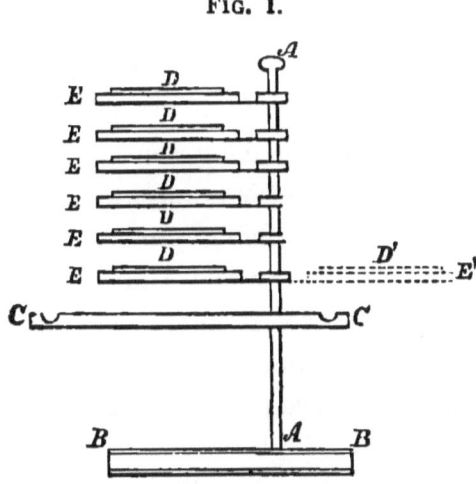

fig. 1. A A is an upright post fixed on a base B B. C C is a square platform of wood round which runs a ditch of water. Above are six nests, D, each lying on a platform E, which could be turned for facility of observation, as shown in the dotted lines D' and E'. Thus the ants had a considerable range, as they could wander as far as the water ditch. The object of having the platform C C larger than the supports of the nests

was that if the ants fell, as often happened, they were within the water boundary, and were able to return home. This plan answered fairly well, and saved space, but it did not quite fulfil my hopes, as the ants were so pugnacious, that I was obliged to be very careful which nests were placed on the same stand.

Of course it was impossible to force the ants into these glass nests. On the other hand, when once the right way is known it is easy to induce them to go in. When I wished to start a new nest I dug one up, and brought home the ants, earth, &c., all together. I then put them over one of my artificial nests, on one of the platforms surrounded by a moat of water. Gradually the outer earth dried up, while that between the two plates of glass, being protected from evaporation, retained its moisture. Under these circumstances the ants found it more suitable to their requirements, and gradually deserted the drier mould outside, which I removed by degrees. In the earth between the plates of glass the ants tunnelled out passages, chambers, &c. (fig. 2), varying in form according to the circumstances and species.

Even between the plates of glass the earth gradually dried up, and I had to supply artificial rain from time to time. Occasionally also I gave them an altogether new nest. They seem, however, to get attached to their old homes, and I have one community which has inhabited the same glass case ever since 1874.

It is hardly necessary to say that the individual

ants belonging to the communities placed on the stands just described, knew their own nests perfectly well.

These nests gave me special facilities for observing the internal economy of ant life. Another main difference between my observations and those of previous naturalists has consisted in the careful record of the actions of individual ants. The most convenient mode of marking them was, I found, either with a small dab of paint on the back, or, in the case of bees or wasps, by snipping off a fragment at the extremity of the wing. This, I need hardly say, from the structure of the wing, gave the insect no pain; in fact, as it is only necessary to remove a minute portion, not sufficient to make any difference in their flight, they seemed scarcely to notice it. I never found any difficulty in painting bees or wasps; if they are given a little honey they become so intent that they quietly allow the paint to be applied. Of course too much must not be put on, and care must be taken not to touch the wings or cover up the spiracles. Ants require somewhat more delicate treatment, but with a little practice they could also be marked without any real difficulty.

No two species of Ants are identical in habits; and, on various accounts, their mode of life is far from easy to unravel. In the first place, most of their time is passed underground: all the education of the young, for instance, is carried on in the dark. Again, ants are essentially gregarious; it is in some cases difficult to keep a few alive by themselves in captivity, and at any rate their habits under such circumstances are entirely altered. If, on the other hand, a whole community is kept, then the greater number introduces a fresh element of difficulty and complexity. Moreover, within the same species, the individuals seem to differ in character, and even the same individual will behave very differently under different circumstances. Although, then, ants have attracted the attention of many of the older naturalists,—Gould, De Geer, Reaumur, Swammerdam, Latreille, Leuwenhoeck, Huber,—and have recently been the object of interesting observations by Frederick Smith, Belt, Moggridge, Bates, Mayr, Emery, Forel, McCook, and others, they still present one of the most promising fields for observation and experiment.

The life of an ant falls into four well-marked periods—those of the egg, of the larva or grub, of the pupa or chrysalis, and of the perfect insect or imago. The eggs are white or yellowish, and somewhat elongated. They are said to hatch about fifteen days after being laid. Those observed by me have taken a month or six weeks.

The larvæ of ants (Pl. V. fig. 3), like those of bees and wasps, are small, white, legless grubs somewhat conical in form, being narrow towards the head. They are carefully tended and fed, being carried about from chamber to chamber by the workers, probably in order to secure the most suitable amount of warmth and moisture. I have observed, also, that they are very often assorted according to age. It is sometimes very curious in my nests to see them arranged in groups according to size, so that they remind one of a school divided into five or six classes.

As regards the length of life of the larvæ, Forel supposed[1] that those of *Tapinoma* matured the quickest, and were full-grown in about six or seven weeks. Some of *Myrmica ruginodis*, however, observed by me, turned into pupæ in less than a month. In other cases the period is much longer. In certain species, *Lasius flavus*, for instance, some of the larvæ live through the winter.

When full grown they turn into pupæ (Pl. V. fig. 4), sometimes naked, sometimes covered with a silken cocoon, constituting the so-called 'ant-eggs.' We do not yet understand why some larvæ spin cocoons, while others remain naked. As a general rule, the species which have not a sting, spin a cocoon, while those which have, are naked. Latreille was the first to observe that in one species (*F. fusca*) the pupæ sometimes spin a cocoon, and sometimes remain naked. The reason for this difference is still quite unknown. After remaining some days in this state they emerge as perfect insects. In many cases, however, they would perish in the attempt, if they were not assisted; and it is very pretty to see the older ants helping them to extricate them-

---

* *Les Fourmis de la Suisse*, p 426.

selves, carefully unfolding their legs and smoothing out the wings, with truly feminine tenderness and delicacy. Our countryman Gould was the first to observe, and the fact has since been fully confirmed by Forel, that the pupæ are unable to emerge from the cocoons without the assistance of the workers. The ants generally remain from three to four weeks in this condition.

In the case of ants, as with other insects which pass through similar metamorphoses, such as bees, wasps, moths, butterflies, flies, and beetles, &c., the larval stage is the period of growth. During the chrysalis stage, though immense changes take place, and the organs of the perfect insect are more or less rapidly developed, no food is taken, and there is no addition to the size or weight.

The imago or perfect insect again takes food, but does not grow. The ant, like all the insects above named, is as large when it emerges from the pupa as it ever will be; excepting, indeed, that the abdomen of the females sometimes increases in size from the development of the eggs.

We have hitherto very little information as to the length of life in ants in the imago, or perfect, state. So far, indeed, as the preparatory stages are concerned, there is little difficulty in approximately ascertaining the facts; namely, that while in summer they take only a few weeks; in some species, as our small yellow meadow ants, the autumn larvæ remain with comparatively little change throughout the winter. It is much more difficult to ascertain the length of life of the perfect insect, on account of their gregarious habits, and the difficulty of recognising individual ants. I have found, however, as we shall presently see, that their life is much longer than has been generally supposed.

It is generally stated in entomological works that the males of ants die almost immediately. No doubt this is generally the case. At the same time, some males of *Myrmica ruginodis*, which I isolated with their mates in August 1876, lived until the following spring; one of them till May 17.

It has also been the general opinion that the females lived about a year. Christ[1] indeed thought they might last three or even four seasons, but this was merely a suggestion, and Forel expressed the general opinion when he said, 'Je suis persuadé qu'en automne il ne reste presque plus que les ouvrières écloses pendant le courant de l'été.' The average life of a queen is also, he thinks, not more than twelve months. I have found, however, that the life of the queens and workers is much longer than had been supposed. I shall give further details in a subsequent chapter, but I may just mention here that I kept a queen of *Formica fusca* from December 1874 till August 1888, when she must have been nearly fifteen years old, and of course may have been more. She

[1] *Naturgeschichte der Insekten.*

attained, therefore, by far the greatest age of any insect on record. I have also some workers which I have had since 1875.

The body of an ant consists of three parts: the head, thorax, and abdomen. The head bears the principal organs of sense, and contains the brain, as the anterior portion of the nervous system may fairly be called. The thorax, supporting the legs and, when they are present, the wings, contains the principal muscles of locomotion. The abdomen contains the stomach and intestines, the organs of reproduction, the sting, &c.

Returning to the head: the antennæ consist of a short spherical basal piece, a long shaft, known as the scape, and a flagellum of from six to seventeen (generally, however, from ten to thirteen) short segments, the apical ones sometimes forming a sort of club. The number of segments is generally different in the males and females.

The eyes are of two kinds. Large compound eyes, one on each side of the head; and ocelli, or so-called simple eyes. The compound eyes consist of many facets. The number differs greatly in different species, and in the different sexes, the males generally having the greatest number. Thus, in *Formica pratensis* there are, according to Forel, in the males about 1,200 in each eye, in the fertile females between 800 and 900, in the workers about 600. Where the workers vary in size

[1] Having reference to the facts stated on page 37, this is a result of great physiological interest.

they differ also in the number of facets. Thus, again following the same authority, the large workers of *Camponotus ligniperdus* have 500, the smaller ones only 450; while in the Harvesting ant (*Atta barbara*) the contrast is even greater, the large specimens having 230, the small ones only from 80 to 90. The ordinary workers have in *Polyergus rufescens* about 400; in *Lasius fuliginosus,* 200; in *Tapinoma erraticum,* 100; in *Plagiolepis pygmæa,* 70 to 80; in *Lasius flavus,* about 80; in *Bothriomyrmex meridionalis,* 55; in *Strongylognathus testaceus, Stenamma Westwoodii,* and *Tetramorium cæspitum,* about 45; in *Pheidole pallidula,* about 30; *Myrmecina Latreillei,* 15; *Solenopsis fugax,* 6 to 9; while in *Ponera contracta* there are only from 1 to 5; in *Eciton* only 1; and in *Typhlopone* the eyes are altogether wanting.

The number of facets seems to increase rather with the size of the species than with the power of vision.

The ocelli are never more than three in number, disposed in a triangle with the apex in front Sometimes the anterior ocellus alone is present. In some species the workers are altogether without ocelli, which, however, are always present in the queens and in the males.

The mouth parts are the labrum, or upper lip; the first pair of jaws or mandibles; the second pair of jaws or maxillæ, which are provided with a pair of palpi, or feelers; and the lower lip, or labium, also bearing a pair of palpi

The thorax is generally considered to consist, as in other insects, of three divisions—the prothorax, mesothorax, and metathorax. I have elsewhere, however, given reasons into which I will not at this moment enter, for considering that the first abdominal segment has in this group coalesced with the thorax. The thorax bears three pairs of legs, consisting of a coxa, trochanter, femur, tibia and tarsus, the latter composed of five segments and terminating in a pair of strong claws.

In the males and females the meso- and metathorax each bear a pair of wings, which, however, are stripped off by the insects themselves soon after the marriage flight.

The workers never possess wings, nor do they show even a rudimentary representative of these organs. Dr. Dewitz has pointed out that the full-grown larvæ of the workers possess well-developed 'imaginal disks,' like those which, in the males and females, develope into the wings. These disks, during the pupal life, gradually become atrophied, until in the perfect insects they are represented only by two strongly chitinised points lying under the large middle thoracic stigmas. No one unacquainted with the original history of these points would ever suspect them to be the rudimentary remnants of ancestral wings.[1]

The thorax also bears three pairs of spiracles, or breathing holes.

[1] *Zeit. f. wiss. Zool.*, vol. xxviii. p. 553.

The abdomen consists of six segments, in the queens and workers, that is to say in the females, and seven in the males. The first segment, as a general rule, in the Formicidæ forms a sort of peduncle (known as the scale or knot) between the metathorax and the remainder of the abdomen. In the Myrmicidæ two segments are thus detached from the rest.

The Poneridæ form, as regards the peduncle, and in some other respects, an intermediate group between the Formicidæ and the Myrmicidæ. The second abdominal segment is contracted posteriorly, but not so much so as to form a distinct knot.

The form of the knot offers in many cases valuable specific characters.

I have sometimes been tempted to correlate the existence of a second knot among the Myrmicidæ with their power of stinging, which is wanting in the Formicidæ. Though the principal mobility of the abdomen is given in the former, as in the latter, by the joint between the metathorax and the knot, still the second segment of the peduncle must increase the flexibility, which would seem to be a special advantage to those species which have a sting. It must indeed be admitted that Œcophylla[1] has a sting, and yet only one knot; but this, of course, does not altogether negative my suggestion, which, however, I only throw out for consideration

[1] *Proc. Linn. Soc.*, vol. v p. 101.

The knot is provided with a pair of spiracles, which are situated, as Forel states, in the front of the segment, and not behind, as supposed by Latreille.

In most entomological works it is stated that the Myrmicidæ have a sting, and that, on the contrary, the Formicidæ do not possess one. The latter family, indeed, possess a rudimentary structure representing the sting, but it seems merely to serve as a support for the poison duct. Dr. Dewitz, who has recently published[1] an interesting memoir on the subject, denies that the sting in Formicidæ is a reduced organ, and considers it rather as in an undeveloped condition. The ancestors of our existing Ants, in his opinion, had a large poison apparatus, with a chitinous support like that now present in Formica, from which the formidable weapons of the bees, wasps, and Myrmicidæ have been gradually developed. I confess that I am rather disposed, on the contrary, to regard the condition of the organ in Formica as a case of retrogression contingent upon disuse.[2] I find it difficult to suppose that organs—so complex, and yet so similar— as the stings of ants, bees, and wasps, should have been developed independently.

Any opinion expressed by M. Dewitz on such a subject is, of course, entitled to much weight; nevertheless there are some general considerations which seem to me conclusive against his view. If the sting

---

[1] *Zeit. f. wiss. Zool.*, vol. xxviii. p. 527.

[2] This view has subsequently been adopted by Dr. Beyer, *Jena Zeit.* 1890.

of Formica represents a hitherto undeveloped organ, then the original ant was stingless, and the present stings of ants have an origin independent of that belonging to the other aculeate Hymenoptera, such as bees and wasps. These organs, however, are so complex, and at the same time so similarly constituted, that they must surely have a common origin. Whether the present sting is derived from a leaf-cutting instrument, such as that from which the sawfly takes its name, I will at present express no opinion. Dr. Dewitz himself regards the rudimentary traces of wings in the larvæ of ants as the remnants of once highly-developed organs; why, then, should he adopt the opposite view with reference to the rudimentary sting? On the whole, I must regard the ancestral ant as having possessed a sting, and consider that the rudimentary condition of that of Formica is due to atrophy, perhaps through disuse.

On the other hand, it is certainly, at first sight, difficult to understand why ants, having once acquired a sting, should allow it to fall into desuetude. There are, however, some considerations which may throw a certain light on the subject. The poison glands are much larger in Formica than in Myrmica. Moreover, some species have the power of ejecting their poison to a considerable distance. In Switzerland, after disturbing a nest of *Formica rufa*, or some nearly allied species, I have found that a hand held as much as 18 inches above the ants was covered with acid. But even when the poison is not thus fired at the enemy from a distance, there are two cases in which the aculeus might be allowed to fall into disuse. Firstly, those species which fight with their mandibles might find it on the whole most convenient to eject the poison (as they do) into the wounds thus created. Secondly, if the poison itself is so intensified in virulence as to act through the skin, a piercing instrument would be of comparatively small advantage. I was amused one day by watching some specimens of the little *Cremastogaster sordidula* and the much larger *Formica cinerea*. The former were feeding on some drops of honey, which the Formicas were anxious to share, but the moment one approached, the little Cremastogasters simply threatened them with the tip of their abdomen, and the Formicas immediately beat a hasty retreat. In this case the comparatively large Formica could certainly have had nothing to fear from physical violence on the part of the little Cremastogaster. Mere contact with the poison, however, appeared to cause them considerable pain, and generally the threat alone was sufficient to cause a retreat.

However this may be, in their modes of fighting, different species of ants have their several peculiarities. Some also are much less military than others. *Myrmecina Latreillii*, for instance, never attack, and scarcely even defend themselves. Their skin is very hard, and they roll themselves into a ball, not defending themselves even if their nest is invaded; to pre-

MODES OF FIGHTING.

vent which they make the entrances small, and often station at each a worker, who uses her head to stop the way. The smell of this species is also, perhaps, a protection. *Tetramorium cæspitum* has the habit of feigning death. This species, however, does not roll itself up, but merely applies its legs and antennæ closely to the body.

Formica rufa, the common Horse ant, attacks in serried masses, seldom sending out detachments, while single ants scarcely ever make individual attacks. They rarely pursue a flying foe, but give no quarter, killing as many enemies as possible, and never hesitating, with this object, to sacrifice themselves for the common good.

Formica sanguinea, on the contrary, at least in their slave-making expeditions, attempt rather to terrify than to kill. Indeed, when invading a nest, they do not attack the flying inhabitants unless these are attempting to carry off pupæ, in which case the *F. sanguineas* force them to abandon the pupæ. When fighting, they attempt to crush their enemies with their mandibles.

Formica exsecta is a delicate, but very active species. They also advance in serried masses, but in close quarters they bite right and left, dancing about to avoid being bitten themselves. When fighting with larger species they spring on to their backs, and then seize them by the neck or by an antenna. They also have the instinct of acting together, three or four seizing an enemy at once, and then pulling different ways, so that she on her part cannot get at any one of her foes. One of them then jumps on her back and cuts, or rather saws, off her head. In battles between this ant and the much larger *F. pratensis*, many of the *F. exsectas* may be seen on the backs of the *F. pratensis*, sawing off their heads from behind.

The species of *Lasius* make up in numbers what they want in strength. Several of them seize an enemy at once, one by each of her legs or antennæ, and when they have once taken hold they will suffer themselves to be cut in pieces rather than leave go.

Polyergus rufescens, the celebrated slave-making or Amazon ant, has a mode of combat almost peculiar to herself. The jaws are very powerful, and pointed. If attacked—if, for instance, another ant seizes her by a leg—she at once takes her enemy's head into her jaws, which generally makes her quit her hold. If she does not, the *Polyergus* closes her mandibles, so that the points pierce the brain of her enemy, paralysing the nervous system. The victim falls in convulsions, setting free her terrible foe. In this manner a comparatively small force of *Polyergus* will fearlessly attack much larger armies of other species, and suffer themselves scarcely any loss.

Under ordinary circumstances an ants' nest, like a beehive, consists of three kinds of individuals: **workers**, or imperfect females (which constitute the great majority), males, and perfect females. There are, however, often several queens in an ants' nest; while, as we all know, there is never more than one queen mother in a hive. The queens of ants are provided with wings, but after a single flight they tear them off, and do not again quit the nest. In addition to the ordinary workers there is in some species a second, or rather a third, form of female. In almost any ants' nest we may see that the workers differ more or less in size. The amount of difference, however, depends upon the species. In *Lasius niger*, the small brown garden ant, the workers are, for instance, much more uniform than in the little yellow meadow ant, or in *Atta barbara* (Pl. II. figs. 1 and 2), where some of them are much more than twice as large as others. But in certain ants there are differences still more remarkable. Thus, in a Mexican species, *Myrmecocystus*,[1] besides the common workers, which have the form of ordinary neuter ants, there are certain others in which the abdomen is swollen into an immense sub-diaphanous sphere. These individuals are very inactive, and principally as living honey-jars. I have described in a subsequent page a species of *Camponotus* (Pl. IV. fig. 1) from Australia, which presents us with the same remarkable phenomenon. In the genus *Pheidole* (Pl. II. figs. 3 and 4), very common in southern Europe, there are also two distinct forms without any intermediate gradations; one with heads of the usual propor-

[1] Wesmael, *Bull. Acad. Roy. Bruxelles*, vol. v. p. 771.

tion, and a second with immense heads provided with very large jaws. This differentiation of certain individuals so as to adapt them to special functions seems to me very remarkable; for it must be remembered that the difference is not one of age or sex. The large-headed individuals are generally supposed to act as soldiers, and the size of the head enables the muscles which move the jaws to be of unusual dimensions; but the little workers are also very pugnacious. Indeed, in some nests of *Pheidole megacephala*, which I had for some time under observation, the small workers were quite as ready to fight as the large ones.

Again, in the genus *Colobopsis* Emery discovered that two ants, then supposed to be different species, and known as *Colobopsis truncata* and *C. fuscipes*, are really only two forms of one species. In this case the entrance to the nest is guarded by the large-headed form, which may therefore fairly be called a soldier.

Savage observed among the Driver Ants, where also there are two kinds of workers, that the large ones arranged themselves on each side of the column formed by the small ones. They acted, he says, evidently the part of guides rather than of guards. At times they place 'their abdomen horizontally on the ground, and laying hold of fixed points with their hind feet (which together thus acted as a fulcrum), elevate the anterior portion of their bodies to the highest point, open wide their jaws, and stretch forth their antennæ, which for the most part were fixed, as if in the act of listening and watching for approaching danger. They would occasionally drop their bodies to the ground again, run off to one side, and fiercely work their jaws and antennæ, as if having detected some strange sounds in the distance. Discerning nothing, they would quickly return to their posts and resume their positions, thus acting as scouts.'[1]

The same thing has been noticed by other naturalists. Bates, for instance, states that in the marching columns of *Eciton drepanophora* the large-headed workers 'all trotted along empty-handed and outside the column, at pretty regular intervals from each other, like subaltern officers in a marching regiment. . . . I did not see them change their position, or take any notice of their small-headed comrades;' and he says that if the column was disturbed they appeared less pugnacious than the others.

In other species, however, of the same genus, *Eciton vastator* and *E. erratica*, which also have two distinct kinds of workers, the ones with large heads do appear to act mainly as soldiers. When a breach is made in one of their covered ways, the small workers set to work to repair the damage, while the large-headed ones issue forth in a menacing manner, rearing themselves up and threatening with their jaws.

In the Sauba Ant of South America (*Æcodoma cephalotes*), the complexity is carried still further;

[1] Rev. T. S. Savage on the 'Habits of the Driver Ants,' *Trans. Ent. Soc.*, vol v. p. 12.

Lund[1] pointed out that there were two different kinds of workers, but Bates has since shown that there are in this species no less than five classes of individuals, namely: 1. Males. 2. Queens. 3. Small ordinary workers (Pl. III. fig. 2). 4. Large workers (Pl. III. fig. 1), with very large hairy heads. 5. Large workers, with large polished heads. Bates never saw either of these two last kinds do any work at all, and was not able to satisfy himself as to their functions. They have also been called soldiers, but this is obviously a misnomer—at least, they are said never to fight. Bates suggests[2] that they may 'serve, in some sort, as passive instruments of protection to the real workers. Their enormously large, hard, and indestructible heads may be of use in protecting them against the attacks of insectivorous animals. They would be, on this view, a kind of *pièces de resistance*, serving as a foil against onslaughts made on the main body of workers.'

This does not, I confess, appear to me a probable explanation of the fact, and on the whole it seems that the true function of these large-headed forms is not yet satisfactorily explained.

The question then arises whether these different kinds of workers are produced from different eggs.

I am disposed to concur with Westwood in the opinion[3] 'that the inhabitants of the nest have the instinct so to modify the circumstances producing this

---

[1] *Ann. des Sci. Nat.* 1831, p. 122.   [2] *Loc. cit.* p. 31.
[3] *Modern Classification of Insects*, vol. ii. p. 225.

state of imperfection, that some neuters shall exhibit characters at variance with those of the common kind.' This, indeed, credits them with a very remarkable instinct, and yet I see no more probable mode of accounting for the facts. Moreover, the exact mode by which the differences are produced is still entirely unknown.

M. Forel, in his excellent work on ants, has pointed out that very young ants devote themselves at first to the care of the larvæ and pupæ, and that they take no share in the defence of the nest or other out-of-door work until they are some days old. This seems natural, because at first their skin is comparatively soft; and it would clearly be undesirable for them to undertake rough work or run into danger until their armour had had time to harden. There are, however, reasons for thinking that the division of labour is carried still further. I do not allude merely to those cases in which there are completely different kinds of workers, but even to the ordinary workers. In *L. flavus*, for instance, it seems probable that the duties of the small workers are somewhat different from those of the large ones, though no such division of labour has yet been detected. I shall have to record some further observations pointing in the same direction.

The nests of ants may be divided into several classes. Some species, such as our common Horse ant (*Formica rufa*), collect large quantities of materials, such as bits of stick, fir leaves, &c., which they heap up into conical masses. Some construct their nests of earth, the cells being partly above, partly below, the natural level. Some are entirely underground, others eat into the trunks of old trees.

In warmer climates the variations are still more numerous. *Formica bispinosa*, of Cayenne, forms its nest of the cottony matter from the capsules of Bombax. Sykes has described[1] a species of *Myrmica* which builds in trees and shrubs, the nest consisting of thin leaves of cow-dung, arranged like tiles on the roof of a house; the upper leaf, however, covering the whole.

In some cases the nests are very extensive. Bates mentions that while he was at Pará an attempt was made to destroy a nest of the Sauba ants by blowing into it the fumes of sulphur, and he saw the smoke issue from a great number of holes, some of them not less than seventy yards apart.

A community of ants must not be confused with an ant hill in the ordinary sense. Very often indeed a community has only one dwelling, and in most species seldom more than three or four. Some, however, form numerous colonies. M. Forel even found a case in which one nest of *F. exsecta* had no less than two hundred colonies, and occupied a circular space with a radius of nearly two hundred yards. Within this area they had exterminated all the other ants, except a few nests of *Tapinoma erraticum*, which survived, thanks to their great agility. In these cases the number of

[1] *Trans. Ent. Soc.*, vol. i.

ants thus associated together must have been enormous. Even in single nests Forel estimates the numbers at from five thousand to half a million.

Ants also make for themselves roads. These are not merely worn by the continued passage of the ants, as was supposed by Christ, but are actually prepared by the ants, rather however by the removal of obstacles, than by any actual construction, which would indeed not be necessary, the weights to be carried being so small. In some cases these roadways are arched over with earth, so as to form covered ways. In others, the ants excavate regular subterranean tunnels, sometimes of considerable length. The Rev. Hamlet Clark even assures us that he observed one in South America, which passed under the river Parahyba at a place where it was as broad as the Thames at London Bridge. I confess, however, that I have my doubts as to this case, for I do not understand how the continuity of the tunnel was ascertained.

The food of ants consists of insects, great numbers of which they destroy; of honey, honeydew, and fruit: indeed, scarcely any animal or sweet substance comes amiss to them. Some species, such, for instance, as the small brown garden ant (*Lasius niger*, Pl. I. fig. 1), ascend bushes in search of aphides. The ant then taps the aphis gently with her antennæ, and the aphis emits a drop of sweet fluid, which the ant drinks. Sometimes the ants even build covered ways up to and over the aphides, which, moreover, they protect from the attacks of other insects. Our English ants do not store up provision for the winter; indeed, their food is not of a nature which would admit of this. I have indeed observed that the small brown ant sometimes carries seeds of the violet into its nest, but for what purpose is not clear. Some of the southern ants, however, lay up stores of grain (*see* Chapter III.).

Ants have many enemies. They themselves, and still more their young, are a favourite food of many animals. They are attacked also by numerous parasites. If a nest of the brown ants is disturbed at any time during the summer, some small flies may probably be seen hovering over the nest, and every now and then making a dash at some particular ant. These flies belong to the genus *Phora*, and to a species hitherto unnamed, which Mr. Verrall has been good enough to describe for me (*see* Appendix). They lay their eggs on the ants, inside which the larvæ live. Other species of the genus are in the same way parasitic on bees. Ants are also sometimes attacked by mites. On one occasion I observed that one of my ants had a mite attached to the underside of its head. The mite, which maintained itself for more than three months in the same position, was almost as large as the head. The ant could not remove it herself. Being a queen, she did not come out of the nest, so that I could not do it for her, and none of her own companions thought of performing this kind office.

In character the different species of ants differ very

much from one another. *F. fusca* (Pl. I. fig. 3), the one which is pre-eminently the 'slave' ant, is, as might be expected, extremely timid; while the nearly allied *F. cinerea* has, on the contrary, a considerable amount of individual audacity. *F. rufa* (Pl. II. fig. 5), the horse ant, is, according to M. Forel, especially characterised by the want of individual initiative, and always moves in troops; he also regards the genus *Formica* as the most brilliant; though others excel it in other respects, as, for instance, in the sharpness of their senses. *F. pratensis* worries its slain enemies; *F. sanguinea* (Pl. I. fig. 6) never does so. The slave-making ant (*P. rufescens*, Pl. I fig. 5) is, perhaps, the bravest of all. If a single individual finds herself surrounded by enemies, she never attempts to fly, as any other ant would, but transfixes her opponents one after another, springing right and left with great agility, till at length she succumbs, overpowered by numbers. *M. scabrinodis* is cowardly and thievish; during wars among the larger species they haunt the battle-fields and devour the dead. *Tetramorium* is said to be very greedy; *Myrmecina* very phlegmatic.

In industry ants are not surpassed even by bees and wasps. They work all day, and in warm weather, if need be, even at night too. I once watched an ant from six in the morning, and she worked without intermission till a quarter to ten at night. I had put her to a saucer containing larvæ, and in this time she carried off no less than a hundred and eighty-seven to the nest. I had another ant, which I employed in my experiments, under continuous observation several days. When I started for London in the morning, and again when I went to bed at night, I used to put her in a small bottle, but the moment she was let out she began to work again. On one occasion I was away from home for a week. On my return I took her out of the bottle, placing her on a little heap of larvæ about three feet from the nest. Under these circumstances I certainly did not expect her to return. However, though she had thus been six days in confinement, the brave little creature immediately picked up a larva, carried it off to the nest, and after half an hour's rest returned for another.

Our countryman Gould noticed[1] certain 'amusements' or 'sportive exercises,' which he had observed among ants. Huber also mentions[2] scenes which he had witnessed on the surface of ant hills, and which, he says, 'I dare not qualify with the title gymnastic, although they bear a close resemblance to scenes of that kind.' The ants raised themselves on their hind legs, caressed one another with their antennæ, engaged in mock combats, and almost seemed to be playing hide and seek. Forel entirely confirms Huber's statements, though he was at first incredulous. He says:[3]—

[1] *An Account of English Ants*, p. 103.
[2] *Nat. Hist. of Ants*, p. 197.
[3] *Loc. cit.*, p. 367.

CLEANLINESS.   29

'Malgré l'exactitude avec laquelle il décrit ce fait, j'avais peine à y croire avant de l'avoir vu moi-même, mais une fourmilière pratensis m'en donna l'exemple à plusieurs reprises lorsque je l'approchai avec précaution. Des ⚥ (*i.e.* workers) se saisissaient par les pattes ou par les mandibules, se roulaient par terre, puis se retachaient, s'entraînaient les unes les autres dans les trous de leur dôme pour en ressortir aussitôt après, etc. Tout cela sans aucun acharnement, sans venin ; il était évident que c'était purement amical. Le moindre souffle de ma part mettait aussitôt fin à ces jeux. J'avoue que ce fait peut paraître imaginaire à qui ne l'a pas vu, quand on pense que l'attrait des sexes ne peut en être cause.'

Bates, also, in the case of *Eciton legionis*, observed behaviour which looked to him 'like simple indulgence in idle amusement, the conclusion,' he says, 'that the ants were engaged merely in play was irresistible.'[1]

Lastly, I may observe that ants are very cleanly animals, and assist one another in this respect. I have often seen them licking one another. Those, moreover, which I painted for facility of recognition were gradually cleaned by their friends.

[1] *Loc. cit.*, vol. ii. p. 362.

CHAPTER II.

ON THE FORMATION AND MAINTENANCE OF NESTS, AND ON THE DIVISION OF LABOUR.

It is remarkable that notwithstanding the researches of so many excellent observers, and though ants' nests swarm in every field and every wood, we did not know how their nests commence.

Three principal modes have been suggested. After the marriage-flight the young queen may either—

1. Join her own or some other old nest;
2. Associate herself with a certain number of workers, and with their assistance commence a new nest; or
3. Found a new nest by herself.

The question can of course only be settled by observation, and the experiments made to determine it had hitherto been indecisive.

Blanchard, indeed, in his work on the 'Metamorphoses of Insects' (I quote from Dr. Duncan's translation, p. 205), says:—'Huber observed a solitary female go down into a small under-ground hole, take off her own wings, and become, as it were, a worker; then she constructed a small nest, laid a few eggs, and brought up the larvæ by acting as mother and nurse at the same time.'

This, however, is not a correct version of what Huber says. His words are :—' I enclosed several females in a vessel full of light humid earth, with which they constructed lodges, where they resided, some singly, others in common. They laid their eggs and took great care of them; and notwithstanding the inconvenience of not being able to vary the temperature of their habitation, they reared some, which became larvæ of a tolerable size, but which soon perished from the effect of my own negligence.'[1]

It will be observed that it was the eggs, not the larvæ, which, according to Huber, these isolated females reared. It is true that he attributes the early and uniform death of the larvæ to his own negligence, but the fact remains that in none of his observations did an isolated female bring her offspring to maturity.

Other entomologists, especially Forel and Ebrard, have repeated the same observations with similar results; and as yet in no single case had an isolated female been known to bring her young to maturity. Forel even thought himself justified in concluding, from his observations and from those of Ebrard, that such a fact could not occur.

Lepeletier de St. Fargeau[2] was of opinion that ants' nests originate in the second mode indicated above, and

---

[1] *Natural History of Ants*, Huber, p. 121.
[2] *Hist. Nat. des Ins. Hyménoptères*, vol. i. p. 143.

it is, indeed, far from improbable that this may occur. No clear case has, however, yet been observed. M. de St. Fargeau himself observes[1] that 'les particularités qui accompagnent la formation première d'une fourmilière sont encore incertaines et mériteraient d'être observées avec soin.'

Under these circumstances I made the following experiments:—

1*a*. I took an old, fertile, queen from a nest of *Lasius flavus*, and put her to another nest of the same species. The workers became very excited and attacked her.

b. I repeated the experiment, with the same result.

c. Do. do. In this case the nest to which the queen was transferred was without a queen; still they would not receive her.

d and *e*. Do. do. do.

I conclude, then, that, at any rate in the case of *L. flavus*, the workers will not adopt an old queen from another nest.

The following observation shows that, at any rate in some cases, isolated queen ants are capable of giving origin to a new community.

On August 14, 1876, I isolated two pairs of *Myrmica ruginodis* which I found flying in my garden. I placed them with damp earth, food, and water, and they continued perfectly healthy through the winter.

---

[1] *Hist. Nat. des Ins. Hyménoptères*, vol. i. p. 144

In April one of the males died, and the second in the middle of May. The first eggs were laid between April 12 and 23. They began to hatch the first week in June, and the first larva turned into a chrysalis on the 27th; a second on the 30th; a third on July 1, when there were also seven larvæ and two eggs. On the 8th there was another egg. On July 8 a fourth larva had turned into a pupa. On July 11 I found there were six eggs, and on the 14th about ten. On the 15th one of the pupæ began to turn brown, and the eggs were about 15 in number. On the 16th a second pupa began to turn brown. On the 21st a fifth larva had turned into a pupa, and there were about 20 eggs. On July 22 the first worker emerged, and a sixth larva had changed. On the 25th I observed the young worker carrying the larvæ about when I looked into the nest; a second worker was coming out. On July 28 a third worker emerged, and a fourth on Aug 5. The eggs appeared to be less numerous, and some had probably been devoured.

This experiment shows that the queens of *Myrmica ruginodis* have the instinct of bringing up larvæ and the power of founding communities. The workers remained about six weeks in the egg, a month in the state of larvæ, and twenty-five to twenty-seven days as pupæ.

Since, however, cases are on record in which communities are known to have existed for many years, it seems clear that fresh queens must be sometimes adopted. I have indeed recorded several experiments in which fertile queens introduced into queenless nests were ruthlessly attacked, and subsequent experiments have always had the same result. Mr. Jenner Fust, however, suggested to me to introduce the queen into the nest, as is done with bees, in a wire cage, and leave her there for two or three days, so that the workers might, as it were, get accustomed to her. Accordingly I procured a queen of *F. fusca* and put her with some honey in a queenless nest, enclosed in a wire cage so that the ants could not get at her. After three days I let her out, but she was at once attacked. Perhaps I ought to have waited a few days longer. On the contrary, Mr. McCook reports a case of the adoption of a fertile queen of *Cremastogaster lineolata* by a colony of the same species:[1]—'The queen,' he says, 'was taken April 16, and on May 14 following was introduced to workers of a nest taken the same day. The queen was alone within an artificial glass formicary, and several workers were introduced. One of these soon found the queen, exhibited much excitement but no hostility, and immediately ran to her sister workers, all of whom were presently clustered upon the queen. As other workers were gradually introduced they joined their comrades, until the body of the queen (who is much larger than the workers) was nearly covered with them. They appeared to be holding on by their mandibles to the delicate hairs upon the female's body, and

---

[1] *Proc. Acad Natural Sciences of Philadelphia*, 1879. 'Note on the Adoption of an Ant-Queen,' by Mr McCook, p. 139.

continually moved their antennæ caressingly. This sort of attention continued until the queen, escorted by workers, disappeared in one of the galleries. She was entirely adopted, and thereafter was often seen moving freely, or attended by guards, about the nest, at times engaged in attending the larvæ and pupæ which had been introduced with the workers of the strange colony. The workers were fresh from their own natural home, and the queen had been in an artificial home for a month.'

In no case, however, when I have put a queen into one of my nests has she been accepted.

Possibly the reason for the difference may be that the ants on which I experimented had been long living in a republic; for, I am informed, that if bees have been long without a queen it is impossible to induce them to accept another.

Moreover, I have found that when I put a queen with a few ants from a strange nest they did not attack her, and by adding others gradually, I succeeded in securing the throne for her.

It is generally stated that among ants the queens only lay eggs. This, however, is not correct.

Denny[1] and Lespès[2] have shown that the workers also are capable of producing eggs; but the latter asserted that these eggs never come to maturity. Forel, however, has proved[3] that this is not the case, but

[1] *Ann. and Mag. Nat. Hist.*, 2nd ser., vol. i.
[2] *Ann. des Sci. Nat.*, 1863.
[3] *Fourmis de la Suisse*, p. 329.

that in some cases, at any rate, the eggs do produce young. Dewitz even maintains [1] that the workers habitually lay eggs, and explains the difference which on this view exists between the workers of ants and those of bees, on the ground that (as he supposes) the majority of ants die in the autumn, so that the eggs laid by the queens alone would not be sufficient to stock the nest in the spring; while among bees the majority survive the winter, and consequently the eggs laid by the queen are sufficient to maintain the numbers of the community. In reply to this argument, it may be observed that among wasps the workers all perish in the autumn, while, on the contrary, among ants I have proved that, at least as regards many species, this is not the case. Moreover, although eggs are frequently laid by workers, this is not so often the case as Dewitz appears to suppose. Forel appears to have only observed it in one or two cases. In my nests the instances were more numerous; and, indeed, I should say that in most nests there were a few fertile workers.

Among bees and wasps also the workers are occasionally fertile; but, so far as our observations go, it is a curious fact that their eggs never produce females, either queens or workers, but always males. The four or five specimens bred by Forel from the eggs of workers were, moreover, all males.

It became therefore an interesting question whether

[1] *Zeit. wiss. Zool.*, vol. xxviii. p. 536.

the same is the case among ants; and my nests have supplied me with some facts bearing on the question. Most of my nests contained queens; and in these it would be impossible, or at least very difficult, to distinguish and follow the comparatively few eggs laid by the workers. Some of my nests, however, contained no queen; and in them therefore all the eggs must have been laid by workers.

One of these was a nest of *Formica cinerea*, which I brought back from Castellamare in November 1875. At that time it contained no eggs or larvæ. In 1876 a few eggs were laid, of which fifteen came to maturity, and were, I believe, all males. In 1877 there were fourteen pupæ, of which twelve came to maturity, and were all males.

Again, in a nest of *Lasius niger*, kept in captivity since July 1875, there were in 1876 about 100 young; and these were, as far as I could ascertain, all males. At any rate there were about 100 males, and I could not find a single young female. In 1877 there were again some pupæ; but owing to an accident none of them came to maturity. In 1878 fifteen came to maturity; and fourteen were males. The other I could not find after it left the pupa skin; but I have no doubt, from the appearance of the pupa, that it was also a male.

Another nest of *Lasius niger*, taken in November 1875, brought in 1878 only one young ant to maturity; and this was a male.

Again, in a nest of *Formica fusca*, taken in 1875, though in 1876 and 1877 eggs were laid and a few arrived at the pupa-state, none came to maturity. They were all, however, either males or queens, and, I have little doubt, were males. In 1878 one came to maturity, and it was a male.

A nest of *F. fusca*, captured in 1876, did not bring up any young in 1877. In 1878 three larvæ came to maturity; and they all proved to be males. Another nest of *F. fusca*, captured in 1877, in 1878 brought only one young one to maturity. This was a male.

In the following year, I again carefully watched my nests, to see what further light they would throw on the subject.

In six of those which contained no queen, eggs were produced, which of course must necessarily have been laid by workers.

The first of these, the nest of *Lasius niger*, which I have watched since July 1875, and which, therefore, is interesting from the great age of the workers, about ten larvæ were hatched, but only four reached the pupa state. Of these one disappeared; the other three I secured, and on examination they all proved to be males. The nest of *Lasius niger*, which has been under observation since November 1875, produced about ten pupæ. Of these I examined seven, all of which I found to be males. The others escaped me. I believe that, having died, they were brought out **and** thrown away.

The nest of *Formica cinerea*, captured at the same time, produced four larvæ, all of which perished before arriving at the pupa stage. The larvæ of males and of queens are much larger than those of workers, and these larvæ were too big to have been those of workers.

In a nest of *Formica fusca*, which I have had under observation since August 1876, three pupæ were produced. They were all males. Another nest of *Formica fusca* produced a single young one, which also was a male.

Lastly, my nest of *Polyergus rufescens*, which M. Forel was so good as to send me in the spring of 1876, in 1879 produced twelve pupæ. Eleven of these turned out to be males. The other one I lost; and I have little doubt that it was brought out and thrown away. It was certainly not a worker. As regards the first three of these pupæ, I omitted to record at the time whether they belonged to the *Polyergus* or to the slaves, though I have little doubt that they belonged to the former species. The last eight, at any rate, were males of *Polyergus*.

Indeed, in all of my queenless nests, males have been produced; and in not a single queenless nest has a worker laid eggs which have produced a female, either a queen or a worker. Perhaps I ought to add that workers are abundantly produced in those of my nests which possess a queen.

While great numbers of workers and males have

come to maturity in my nests, with one exception not a single queen has been produced.

This was in a nest of *Formica fusca*, in which five queens came to maturity. The nest (which, I need hardly say, possessed a queen) had been under observation since April 1879, and the eggs therefore must have been laid in captivity. The nest had been richly supplied with animal food, which may possibly account for the fact.

It is known that bees, by difference of food, &c., possess the power of obtaining at will from the same eggs either queens or ordinary workers. Mr. Dewitz,[1] however, is of opinion that among ants, on the contrary, the queens and workers are produced from different kinds of eggs. He remarks that it is very difficult to understand how the instinct, if it is to be called instinct, which would enable the working ants to make this difference can have arisen. This is no doubt true; but it seems to me quite as difficult to understand how the queens, which must have originally laid only queen eggs and male eggs, can have come to produce another class. Moreover, however great the difficulty may be to understand how the ants can have learnt to produce queens and workers from one kind of egg, the same difficulty exists almost to the same extent in bees, which, as Mr. Dewitz admits, do possess the power. Moreover, it seems to me very unlikely that the result is produced in one way in the case of

[1] *Zeit. für wiss. Zool.* 1878, p. 101

bees, and in another in that of ants. It is also a strong argument that in none of my nests, though thousands of workers and males have been produced, have I ever observed a queen to be so until this year. On the whole, then, though I differ from so excellent a naturalist with much hesitation, I cannot but think that ants, like bees, possess the power of developing a given egg into either a queen or a worker.

I have already mentioned that the previous views as to the duration of life of ants turn out to be quite erroneous. It was the general opinion that they lived for a single year. Two of my queen ants lived, the one nearly fourteen, the other nearly fifteen years, viz., from December 1874 to July 1887 and August 1888 respectively. During the whole time they enjoyed perfect health, and every year have laid eggs producing workers, a fact which suggests physiological conclusions of great interest.

I have, moreover, little doubt that some of the workers now in this nest were among those originally captured, the mortality after the first few weeks having been but small. This, however, I cannot prove.

A nest of *F. sanguinea*, which M. Forel kindly forwarded to me on September 12, 1875 (but which contained no queen), gradually diminished in numbers, until in February 1879 it was reduced to two *F. sanguineas* and one slave. The latter died in February 1880. One of the two mistresses died between May 10 and May 16, 1880, and the other only survived her a few days, dying between the 16th and 20th. These two ants, therefore, must have been five years old at least. It is certainly curious that they should, after living so long, have died within ten days of one another. There was nothing, as far as I could see, in the state of the nest or the weather to account for this, and they were well supplied with food; yet I hardly venture to suggest that the survivor pined away for the loss of her companion.

Some workers of *F. cinerea* lived in one of my nests from November 1875 to July 1881.

In a nest of *F. fusca*, which I brought in on June 6, 1875, and in one of *Lasius niger* brought in on July 25, 1875, there were no queens; and, as already mentioned, no workers have been produced. Those now living (December 1881) are therefore the original ones, and they must be more than six years old.

The duration of life in ants is therefore much greater than has been hitherto supposed.

Though I lose many ants from accidents, especially in summer, in winter there are very few deaths.

I have given the following figure (fig. 2), which represents a typical nest belonging to *Lasius niger*, because it is a good instance of the mode in which my ants excavated chambers and galleries for themselves, and seems to show some ideas of strategy. The nest is, as usual, between two plates of glass, the outer border is a framework of wood, and the shaded part

ARRANGEMENT OF A NEST.

Fig. 2.

Ground-plan of a typical nest of *Lasius niger*, reduced. *a*, narrow doorway; *b*, hall; *c*, vestibule; *d*, main chamber; *e*, inner sanctum; *f, f, f, f*, narrow entrance passages to sanctum; *g, g*, special pillars

represents garden mould, which the ants have themselves excavated, as shown in the figure. For the small doorway (*a*), indeed, I am myself responsible. I generally made the doorways of my nests narrow, so as to check evaporation and keep the nests from becoming too dry. It will be observed, however, that behind the hall (*b*) the entrance contracts, and is still further protected by a pillar of earth, which leaves on either side a narrow passage which a single ant could easily guard, or which might be quickly blocked up. Behind this is an irregular vestibule (*c*), contracted again behind into a narrow passage, which is followed by another, this latter opening into the main chamber (*d*). In this chamber several pillars of earth are left, almost as if to support the roof. Behind the main chamber is an inner sanctum divided into three chambers, and to which access is obtained through narrow entrances (*f, f, f, f*). Most of the pillars in the main chamber are irregular in outline, but two of them (*g, g*) were regular ovals, and round each, for a distance about as long as the body of an ant, the glass had been most carefully cleaned. This was so marked, and the edge of the cleaned portion was so distinct, that it is impossible not to suppose that the ants must have had some object in this proceeding, though I am unable to suggest any explanation of it.

I have already mentioned (*ante*, p. 23), that there is evidence of some division of labour among ants. Where, indeed, there are different kinds of workers, this is self-evident, but even in species where the workers are all of one type, something of the same kind appears to occur.

In the autumn of 1875 I noticed an ant belonging to one of my nests of *F. fusca* out feeding alone. The next day the same ant was again out by herself, and for some weeks no other ant, so far as I observed, came out to the food. I did not, however, watch her with sufficient regularity. In the winter of 1876, therefore, I kept two nests under close observation, having arranged with my daughters and their governess, Miss Wendland (most conscientious observers), that one of us should look at them once an hour during the day. One of the nests contained about 200 individuals of *F. fusca*, the other was a nest of *P. rufescens* with the usual slaves, about 400 in number. The mistresses themselves never came out for food, leaving all this to the slaves.

We began watching on November 1, but did not keep an hourly register till the 20th, after which date the results are given in the following tables (*see* Appendix). Table No. 1 relates to the nest of *F. fusca*, and the ants are denoted by numbers. The hours at which we omitted to record an observation are left blank; when no ant was at the honey, the square is marked with an 0. An ant, marked in my register as No. 3, was at the time when we began observing acting as feeder to the community.

The only cases in which other ants came to the honey were at 2 P.M. on November 22, when another ant came out, whom we registered as No. 4, another on the 28th, registered as No. 5. Other ants came out occasionally, but not one came to the honey (except the above mentioned) from November 28 till January 3, when another (whom we registered as No. 6) began feeding. After this a friend visited the honey once on the 4th, once on the 11th, and again on the 15th, when she was registered as No. 7.

Table No. 2 is constructed in the same way, but refers to the nest of *Polyergus*. The feeders in this case were, at the beginning of the experiment, registered as Nos. 5, 6, and 7. On November 22 a friend, registered as No. 8, came to the honey, and again on December 11; but with these two exceptions the whole of the supplies were carried in by Nos. 5 and 6, with a little help from No. 7.

Thinking now it might be alleged that possibly these were merely unusually active or greedy individuals, I imprisoned No. 6 when she came out to feed on the 5th. As will be seen from the table, no other ant had been out to the honey for some days; and it could therefore hardly be accidental that on that very evening another ant (then registered as No. 9) came out for food. This ant, as will be seen from the table, then took the place of No. 6, and (No. 5 being imprisoned on January 11) took in all the supplies, again with a little help from No. 7. So matters continued till the 17th, when I imprisoned No. 9, and then again, *i.e.* on the 19th, another ant (No. 10) came out for the food, aided, on and after the 22nd, by another, No. 11. This seems to me very curious. From November 1 to January 5, with two or three casual exceptions, the whole of the supplies were carried in by three ants, one of whom, however, did comparatively little. The other two were imprisoned, and then, but not till then, a fresh ant appears on the scene. She carried in the food for a week; and then, she being imprisoned, two others undertook the task. On the other hand, in Nest 1, where the first foragers were not imprisoned, they continued during the whole time to carry in the necessary supplies.

The facts therefore certainly seem to indicate that certain ants are told off as foragers, and that during winter, when little food is required, two or three are sufficient to provide it.

I have, indeed, no reason to suppose that in our English ants any particular individuals are specially adapted to serve as receptacles of food. M. Wesmael, however, has described [1] a remarkable genus (*Myrmecocystus mexicanus*), brought by M. de Normann from Mexico, in which certain individuals in each nest serve as animated honey-pots. To them the foragers bring their supplies, and their whole duty seems to be to receive the honey, retain it, and redistribute it when required. Their abdomen becomes enormously distended, the intersegmental membranes being so much extended that

[1] *Bull. de l'Acad. des Sci. de Bruxelles*, vol. v. p. 771.

the chitinous segments which alone are visible externally in ordinary ants seem like small brown transverse bars. The account of these most curious insects given by MM. de Normann and Wesmael has been fully confirmed by subsequent observers; as, for instance, by Lucas,[1] Saunders,[2] Edwards,[3] Blake,[4] Loew,[5] and McCook.[6]

On one very important point, however, M. Wesmael was in error; he states that the abdomen of these abnormal individuals 'ne contient aucun organe; ou plutôt, il n'est lui-même qu'un vaste sac stomacal.' Blake even asserts that 'the intestine of the insect is not continued beyond the thorax,' which must surely be a misprint; and also that there is no connexion between the stomach and the intestine! These statements, however, are entirely erroneous; and, as M. Forel has shown, the abdomen does really contain the usual organs, which, however, are very easily overlooked by the side of the gigantic crop.

I have therefore been much interested in receiving a second species of ant, which has been sent me by Mr. Waller, in which a similar habit has been evolved and a similar modification has been produced. The two species, however, are very distinct, belonging to totally

[1] *Ann. Soc. Ent. de France*, v. p. 111.
[2] *Canadian Entomologist*, vol. vii. p. 12.
[3] *Proc. California Academy*, 1873.
[4] *Ibid.*, 1874.
[5] *American Nat.*, viii. 1874.
[6] *The Honey Ants.*

different genera; and the former is a native of Mexico, while the one now described comes from Adelaide in Australia. The two species, therefore, cannot be descended one from the other; and the conclusion seems inevitable that the modification has originated independently in the two species.

It is interesting that, although these specimens apparently never leave the nest, and have little use therefore for legs, mandibles, &c., the modifications which they have undergone seem almost confined to the abdominal portion of the digestive organs. The head and thorax, antennæ, jaws, legs, &c. differ but little from those of ordinary ants.

CHAPTER III.

ON THE RELATION OF ANTS TO PLANTS.

It is now generally admitted that the form and colour, the scent and honey of flowers, are mainly due to the unconscious agency of insects, and especially of bees Ants have not exercised so great an influence over the vegetable kingdom, nevertheless they have by no means been without effect.

The great object of the beauty, scent, and honey of flowers, is to secure cross fertilisation; but for this purpose winged insects are almost necessary, because they fly readily from one plant to another, and generally confine themselves for a certain time to the same species. Creeping insects, on the other hand, naturally would pass from one flower to another on the same plant; and as Mr. Darwin has shown, it is desirable that the pollen should be brought from a different plant altogether. Moreover, when ants quit a plant, they naturally creep up another close by, without any regard to species. Hence, even to small flowers, such as many crucifers, composites, saxifragés, &c., which, as far as size is concerned, might well be fertilised by ants, the visits of flying insects are much more advantageous. Moreover, if larger flowers were visited by ants, not only would they deprive the flowers of their honey without fulfilling any useful function in return, but they would probably prevent the really useful visits of bees. If you touch an ant with a needle or a bristle, she is almost sure to seize it in her jaws; and if bees, when visiting any particular plant, were liable to have the delicate tip of their proboscis seized on by the horny jaws of an ant, we may be sure that such a species of plant would soon cease to be visited. On the other hand, we know how fond ants are of honey, and how zealously and unremittingly they search for food. How is it then that they do not anticipate the bees, and secure the honey for themselves? This is guarded against in several ways.

Belt appears to have been the first naturalist to call attention to this interesting subject.

'Many flowers,' he says,[1] 'have contrivances for preventing useless insects from obtaining access to the nectaries.

. . . .

'Great attention has of late years been paid by naturalists to the wonderful contrivances amongst flowers to secure cross fertilisation, but the structure of many cannot, I believe, be understood, unless we take into consideration not only the beautiful adaptations for securing the services of the proper insect or

---

[1] *The Naturalist in Nicaragua.* By Thos. Belt, F.G.S., pp. 131 and 133.

bird, but also the contrivances for preventing insects that would not be useful from obtaining access to the nectar. Thus the immense length of the *Angræcum sesquipedale* of Madagascar might, perhaps, have been more easily explained by Mr. Wallace, if this important purpose had been taken into account.'

Kerner has since published a very interesting work,[1] especially devoted to the subject, which has been translated into English by Dr. Ogle.

In aquatic plants, of course, the access of ants is precluded by the isolation in water. Nay, even many land plants have secured to themselves the same advantage, the leaves forming a cup round the stem. Some species have such a leaf-cup at each joint, in others there is only a single basin, formed by the rosette of radical leaves. In these receptacles rain and dew not only collect, but are retained for a considerable time. In our own country *Dipsacus sylvestris* (the common teazle) is the best marked instance of this mode of protection, though it is possible that these cups serve another purpose, and form, as suggested by Francis Darwin, traps in which insects are caught, and in which they are dissolved by the contained fluid, so as to serve as food for the plant. However this may be, the basins are generally found to contain water, even if no rain has fallen for some days, and must, therefore, serve to prevent the access of ants.

The next mode of protection is by means of slippery

[1] Kerner: *Flowers and their Unbidden Guests*.

surfaces. In this case, also, the leaves often form a collar round the stem, with curved surfaces over which ants cannot climb. 'I have assured myself,' says Kerner, 'not only by observation, but by experiment, that wingless insects, and notably ants, find it impossible to mount upwards over such leaves as these. The little creatures run up the stem, and may even not unfrequently traverse the under surface of the leaves, if not too smooth; but the reflexed and slippery margin is more than the best climbers among them can get over, and if they attempt it they invariably fall to the ground. There is no necessity for the lamina of the leaf to be very broad; even narrow leaves, as, for instance, those of *Gentiana firma*, are enough for the purpose, supposing, of course, that the margin is bent backwards in the way described.'

Of this mode of protection the cyclamen and snowdrop offer familiar examples. In vain do ants attempt to obtain access to such flowers, the curved surfaces baffle them; when they come to the edge they inevitably drop off to the ground again. In fact, these pendulous flowers protect the honey as effectually from the access of ants, as the hanging nests of the weaver and other birds protect their eggs and young from the attacks of reptiles.

In a third series of plants the access of creeping insects is impeded or altogether prevented by certain parts of the flower being crowded together so as to leave either a very narrow passage or none at all. Thus the *Antirrhinum*, or Snapdragon, is completely closed, and only a somewhat powerful insect can force its way in. The flower is in fact a strong box, of which the Humble-bee only has the key. The Linarias are another case of this kind. The Campanulas, again, are open flowers, but the stamens are swollen at the base, and in close contact with one another, so that they form the lid of a hollow box in which the honey is secreted. In some species the same object is effected by the stamens being crowded together, as in some of the white ranunculuses of the Alps. In other cases, the flower forms a narrow tube, still further protected by the presence of hairs, sometimes scattered, sometimes, as in the white dead nettle, forming a row.

In others, as in some species of Narcissus, Primula, Pedicularis, &c., the tube itself is so narrow that even an ant could not force its way down.

In others, again, as in some of the Gentians, the opening of the tube is protected by the swollen head of the pistil.

In others, as in clover, lotus, and many other *Leguminosæ*, the ovary and the stamens, which cling round the ovary in a closely-fitting tube, fill up almost the whole space between the petals, leaving only a very narrow tube.

Lastly, in some, as in *Geranium robertianum*, *Linum catharticum*, &c., the main tube itself is divided by ridges into several secondary ones.

In still more numerous species the access of ants and other creeping insects is prevented by the presence of spines or hairs, which constitute a veritable *chevaux de frise*. Often these hairs are placed on the flowers themselves, as in some verbenas and gentians. Sometimes the whole plant is more or less hairy, and it will be observed that the hairs of plants have a great tendency to point downwards, which of course constitutes them a more efficacious barrier.

In another class of cases access to the flowers is prevented by viscid secretions. Everyone who has any acquaintance with botany knows how many species bear the specific name of 'Viscosa' or 'Glutinosa.' We have, for instance, *Bartsia viscosa, Robinia viscosa, Linum viscosum, Euphrasia viscosa, Silene viscosa, Dianthus viscidus, Senecio viscosus, Holosteum glutinosum,* &c. Even those who have never opened a botanical work must have noticed how many plants are more or less sticky. Why is this? What do the plants gain by this peculiarity? The answer probably is, at any rate in most cases, that creeping insects are thus kept from the flowers. The viscid substance is found most frequently and abundantly on the peduncles immediately below the blossoms, or even on the blossoms themselves. In *Epimedium alpinum,* for instance, the leaves and lower parts of the stem are smooth, while the peduncles are covered with glandular, viscid hairs. The number of small insects which are limed and perish on such plants is very considerable. Kerner counted sixty-four small insects on one inflorescence of *Lychnis viscosa*. In other species the flower is viscid; as, for instance, in the gooseberry, *Linnæa borealis*, *Plumbago Europæa*, &c.

Polygonum amphibium is a very interesting case. The small rosy flowers are richly supplied with honey; but from the structure of the flower, it would not be fertilised by creeping insects. As its name indicates, this plant grows sometimes on land, sometimes in water. Those individuals, however, which grow on dry land are covered by innumerable glandular viscid hairs, which constitute an effectual protection. On the other hand, the individuals which grow in water are protected by their situation. To them the glandular hairs would be useless, and in fact on such specimens they are not developed.

In most of the cases hitherto mentioned the viscid substance is secreted by glandular hairs, but in others it is discharged by the ordinary cells of the surface. Kerner is even of opinion that the milky juice of certain plants—for instance, of some species of *Lactuca* (lettuce)—answers the same purpose. He placed several kinds of ants on these plants, and was surprised to find that their sharp claws cut through the delicate epidermis; while through the minute clefts thus made the milky juice quickly exuded, by which the ants were soon glued down. Kerner is even disposed to suggest that the nectaries which occur on certain leaves are a means of protection against the unwelcome, because unprofitable, visits of creeping insects, by diverting them from the flowers.

Thus, then, though ants have not influenced the present condition of the vegetable kingdom to the same extent as bees, they have also had a very considerable effect upon it in various ways.

Our European ants do not strip plants of their leaves. In the tropics, on the contrary, some species do much damage in this manner.

Bates considers[1] that the leaves are used 'to thatch the domes which cover the entrances to their subterranean dwellings, thereby protecting them from the rains.' Belt, on the other hand, maintains that they are torn up into minute fragments, so as to form a flocculent mass, which serves as a bed for mushrooms; the ants are, in fact, he says, 'mushroom growers and eaters.'[2]

Some trees are protected by one species of ants from others. A species of Acacia, described by Belt, bears hollow thorns, while each leaflet produces honey in a crater-formed gland at the base, as well as a small, sweet, pear-shaped body at the tip. In consequence, it is inhabited by myriads of a small ant, which nests in the hollow thorns, and thus finds meat, drink, and lodging all provided for it. These ants are continually roaming over the plant; and constitute a most efficient bodyguard, not only driving off the leaf-cutting ants, but, in Belt's opinion, rendering the leaves less liable to be

---

[1] *Loc. cit.*, v. i p. 26.
[2] *Loc. cit.*, p. 79. This view has since been confirmed by Schimper, *Bot. Mitt. aus den Tropen.* Nr. 6.

eaten by herbivorous mammalia. Delpino mentions that on one occasion he was gathering a flower of *Clerodendrum fragrans*, when he was himself 'suddenly attacked by a whole army of small ants.'[1]

Moseley has also called attention[2] to the relations which have grown up between ants and two 'curious epiphytes, *Myrmecodia armata* and *Hydnophytum formicarum*. Both plants are associated in their growth with certain species of ants. As soon as the young plants develop a stem, the ants gnaw at the base of this, and the irritation produced causes the stem to swell; the ants continuing to irritate and excavate the swelling, it assumes a globular form, and may become even larger than a man's head.

'The globular mass contains within a labyrinth of chambers and passages, which are occupied by the ants as their nest. The walls of these chambers and the whole mass of the inflated stem retain their vitality and thrive, continuing to increase in size with growth. From the surface of the rounded mass are given off small twigs, bearing the leaves and flowers.

'It appears that this curious gall-like tumour on the stem has become a normal condition of the plants, which cannot thrive without the ants. In *Myrmecodia armata* the globular mass is covered with spine-like excrescences. The trees I referred to at Amboina had these curious spine-covered masses perched in every

---

[1] *Scientific Lectures*, p. 23.
[2] *Notes by a Naturalist on the 'Challenger,'* p. 389.

fork, and with them also smooth surfaced masses of a species of *Hydnophytum*.'

There are, of course, many cases in which the action of ants is very beneficial to plants. They kill off a great number of small caterpillars and other insects. Forel found in one large nest that more than twenty-eight dead insects were brought in per minute; which would give during the period of greatest energy more than 100,000 insects destroyed in a day by the inhabitants of one nest alone.

Our English hunting ants generally forage alone. In warmer countries, however, they hunt in packs, or even armies.

As already mentioned, none of our northern ants store up grain, and hence there has been much discussion as to the well-known passage of Solomon. I have indeed observed that the small brown ants, *Lasius niger*, sometimes carry seeds of the violet into their nests, but for what purpose is not clear. It is, however, now a well-established fact that more than one species of southern ants do collect seeds of various kinds. The fact, of course, has long been known in those regions.

Indeed, the quantity of grain thus stored up is sometimes so considerable, that in the 'Mischna,' rules are laid down with reference to it; and various commentators, including the celebrated Maimonides, have discussed at length the question whether such grain belonged to the owner of the land, or might be taken by gleaners—giving the latter the benefit of the doubt. They do not appear to have considered the rights of the ants.

Hope[1] has called attention to the fact that Meer Hassan Ali, in his 'History of the Mussulmans,' expressly mentions it. 'More industrious little creatures,' he says, 'cannot exist than the small red ants, which are so abundant in India. I have watched them at their labours for hours, without tiring. They are so small, that from eight to twelve in number labour with great difficulty to convey a grain of wheat or barley, yet these are not more than half the size of a grain of English wheat. I have known them to carry one of these grains to their nest, at a distance from 600 to 1,000 yards. They travel in two distinct lines over rough or smooth ground, as it may happen, even up and down steps, at one regular pace. The returning unladen ants invariably salute the burthened ones, who are making their way to the general storehouse; but it is done so promptly, that the line is neither broken nor their progress impeded by the salutation.'

Sykes, in his account of an Indian ant, *Pheidole providens*,[2] appears to have been the first of modern scientific authors to confirm the statements of Solomon. He states that the above-named species collects large stores of grass seeds, on which it subsists from February

---

[1] *Trans. Ent. Soc.* 1840, p. 213.
[2] *Ibid.* 1836, p. 99. Dr. Lincecum has also made a similar observation.

to October. On one occasion he even observed the ants bringing up their stores of grain to dry them after the closing thunderstorms of the monsoon; an observation which has been since confirmed by other naturalists.

It is now known that harvesting ants occur in the warmer part of Europe, where their habits have been observed with care, especially by Moggridge and Lespès. It does not yet seem quite clear in what manner the ants prevent the grains from germinating. Moggridge found that if the ants were prevented from entering the granaries, the seeds began to sprout, and that this was also the case in deserted granaries. It would appear therefore that the power of germination was not destroyed.

On the other hand, Lespès confirms the statement long ago made by Pliny that the ants gnaw off the radicle, while Forel asserts that *Atta structor* allows the seeds in its granaries to commence the process of germination for the sake of the sugar.

A Texan ant, *Pogonomymex barbatus*, is also a harvesting species, storing up especially the grains of *Aristida oligantha*, the so-called 'ant rice,' and of a grass, *Buchlæ dactyloides*. These ants clear disks, ten or twelve feet in diameter, round the entrance to their nest, a work of no small labour in the rich soil, and under the hot sun, of Texas. I say 'clear disks,' but some, though not all, of these disks are occupied, especially round the edge, by a growth of ant rice. These ants were first noticed by Mr. Buckley,[1] and their habits were some time afterwards described in more detail by Dr. Lincecum,[2] who maintained not only that the ground was carefully cleared of all other species of plants, but that this grass was intentionally cultivated by the ants. Mr. McCook,[3] by whom this subject has been recently studied, fully confirms Dr. Lincecum that the disks are kept carefully clean, that the ant rice alone is permitted to grow on them, and that the produce of this crop is carefully harvested; but he thinks that the ant rice sows itself, and is not actually cultivated by the ants. I have myself observed in Algeria, that certain species of plants are allowed by the ants to grow on their nests.

[1] *Proc. Acad. Nat. Sci. Philadelphia*, 1860.
[2] *Linnean Journal*, 1861, p. 29.
[3] *The Nat. Hist. of the Agricultural Ants of Texas*, p. 38.

CHAPTER IV.

ON THE RELATIONS OF ANTS TO OTHER ANIMALS.

The relations existing between ants and other animals are even more interesting than their relations with plants. As a general rule, not, however, without many remarkable exceptions, they may be said to be those of deadly hostility.

Though honey is the principal food of my ants, they are very fond of meat, and in their wild state ants destroy large numbers of other insects. Our English ants generally go out hunting alone, but many of the species living in hotter climates hunt in packs, or even in armies.

Savage has given [1] a graphic account of the 'Driver' ants (*Anomma arcens West.*) of West Africa. They keep down, he says, 'the more rapid increase of noxious insects and smaller reptiles; consume much dead animal matter, which is constantly occurring, decaying, becoming offensive, and thus vitiating the atmosphere, and which is by no means the least important in the Torrid Zone, often compelling the inhabitants to keep

[1] 'On the Habits of the Driver Ants,' *Trans. Ent. Soc.* 1847 p. 1.

their dwellings, towns, and their vicinity in a state of comparative cleanliness. The dread of them is upon every living thing. . . . .

'Their entrance into a house is soon known by the simultaneous and universal movement of rats, mice, lizards, Blapsidæ, Blattidæ, and of the numerous vermin that infest our dwellings. Not being agreed, they cannot dwell together, which modifies in a good measure the severity of the Drivers' habits, and renders their visits sometimes (though very seldom in my view) desirable. . . . .

'They move over the house with a good degree of order, unless disturbed, occasionally spreading abroad, ransacking one point after another, till, either having found something desirable, they collect upon it, when they may be destroyed *en masse* by hot water. . . . .

'When they are fairly in, we give up the house, and try to await with patience their pleasure, thankful, indeed, if permitted to remain within the narrow limits of our beds or chairs.'

These ants will soon destroy even the largest animal if it is confined. In one case Savage saw them kill near his house a snake four feet long. Indeed, it is said that they have been known to destroy the great python, when gorged with food and powerless. The natives even believe that the python, after crushing its victim, does not venture to swallow it, until it has made a search, and is satisfied that there are no Drivers in the vicinity! It is very remarkable that these hunting ants are blind. They emerge, however, principally by night, and like some of the blind hunting ants of Brazil (*Eciton vastator* and *E. erratica*), well described by Bates,[1] prefer to move under covered galleries, which they construct rapidly as they advance. 'The column of foragers pushes forward step by step, under the protection of these covered passages, through the thickets, and on reaching a rotting log, or other promising hunting ground, pour into the crevices in search of booty.'

The marauding troops of Ecitons may, in some cases, be described as armies. 'Wherever they move,' says Bates,[2] 'the whole animal world is set in commotion, and every creature tries to get out of their way. But it is especially the various tribes of wingless insects that have cause for fear, such as heavy-bodied spiders, ants of other species, maggots, caterpillars, larvæ of cockroaches, and so forth, all of which live under fallen leaves or in decaying wood. The Ecitons do not mount very high on trees, and therefore the nestlings of birds are not much incommoded by them. The mode of operation of these armies, which I ascertained, only after long-continued observation, is as follows: The main column, from four to six deep, moves forward in a given direction, clearing the ground of all animal matter dead or alive, and throwing off, here and there, a thinner column to forage for a short time on the

---

[1] *The Naturalist on the River Amazon*, vol. ii. p. 364.
[2] *Ibid.*, p. 358.

flanks of the main army, and re-enter it again after their task is accomplished. If some very rich place be encountered anywhere near the line of march—for example, a mass of rotten wood abounding in insect larvæ, a delay takes place, and a very strong force of ants is concentrated upon it.'

Belt, also, has given [1] an excellent account of these Ecitons. He observed that spiders were peculiarly intelligent in escaping them, making off several yards in advance; and not like cockroaches and other stupider insects, taking shelter in the first hiding-place, where they were almost sure to be detected. The only chance of safety was either to run right away or to stand still. He once saw a Harvestman (*Phalangium*) standing in the midst of an army of ants with the greatest circumspection and coolness, lifting its long legs one after the other. Sometimes as many as five out of the eight would be in the air at once, but it always found three or four spots free from ants, on which it could safely place its feet. On another occasion, Belt observed a green leaf-like locust, which remained perfectly still, allowing the ants to run over it. This they did, but seem to have been quite deceived by its appearance and immobility, apparently taking it for a leaf.

In other cases, insects mimic ants, and thus escape attack or are able to stalk their prey. Belt mentions a spider which in its form, colour, and movements so

---

[1] *The Naturalist in Nicaragua*, p. 17.

much resembled an ant, that he was himself for some time deceived.

Nor are ants without their enemies. We all know how fond birds are of their larvæ and pupæ. They have also numerous parasites. I have already alluded to the mites which are often found in ants' nests. These are of several kinds; one of them, not uncommon in the nests of *Lasius flavus,* turned out to be a new species, and has been described for me by Mr. Michael (*see* Appendix).

Certain species of Diptera, belonging to the family Phoridæ, are also parasitic on ants. As already mentioned, I forwarded specimens to Mr. Verrall, who finds that some of them are a new species of the genus Phora, and that among them is also the type of a new genus, which he proposes to call Platyphora, doing me the honour of naming the species after me. I subjoin his description in the appendix.

But the social and friendly relations which exist between ants and other animals are of a more complex and much more interesting character.

It has long been known that ants derive a very important part of their sustenance from the sweet juice excreted by aphides. These insects, in fact, as has been over and over again observed, are the cows of the ants; in the words of Linnæus, 'Aphis formicarum vacca.' A good account of the relations existing between ants and aphides was given more than a hundred years ago by the Abbe Boisier de Sauvages.[1]

Nor are the aphides the only insects which serve as cows to the ants. Various species of Coccidæ, Cercopis, Centrotus, Membracis, &c., are utilised in the same manner. H. Edwards[2] and M'Cook[3] have observed ants licking the larva of a butterfly, *Lycæna pseudargiolus*.

The different species of ants utilise different species of aphis. The common brown garden ant (*Lasius niger*) devotes itself principally to aphides which frequent twigs and leaves; *Lasius brunneus*, to the aphides which live on the bark of trees; while the little yellow ant (*Lasius flavus*) keeps flocks and herds of the root-feeding aphides.

In fact, to this difference of habit the difference of colour is perhaps due. The Baltic amber contains among the remains of many other insects a species of ant intermediate between our small brown garden ants and the little yellow meadow ants. This is possibly the stock from which these and other allied species are descended. One is tempted to suggest that the brown species which live so much in the open air, and climb up trees and bushes, have retained and even deepened their dark colour; while others, such as *Lasius flavus*,

---

[1] *Observations sur l'origine du miel*, par l'Abbé Boisier de Sauvages, *Jour. de Physique*, vol. i. p. 187.

[2] *Canadian Entomologist*, January 1878.

[3] *The Mound-making Ants of the Alleghenies*, p. 289.

the yellow meadow ant, which lives almost entirely below ground, has become much paler.

The ants may be said almost literally to milk the aphides; for, as Darwin and others have shown, the aphides generally retain the secretion until the ants are ready to receive it. The ants stroke and caress the aphides with their antennæ, and the aphides then emit the sweet secretion.

As the honey of the aphides is more or less sticky, it is probably an advantage to the aphis that it should be removed. Nor is this the only service which ants render to them. They protect them from the attacks of enemies; and not unfrequently even build cowsheds of earth over them. The yellow ants collect the root-feeding species in their nests, and tend them as carefully as their own young. But this is not all. The ants not only guard the mature aphides, which are useful; but also the eggs of the aphides, which of course, until they come to maturity, are quite useless. These eggs were first observed by our countryman Gould, whose excellent little work on ants [1] has hardly received the attention it deserves. In this case, however, he fell into error. He states that 'the queen ant' [he is speaking of *Lasius flavus*] 'lays three different sorts of eggs, the slave, female, and neutral. The two first are deposited in the spring, the last in July and part of August; or, if the summer be extremely favourable,

[1] *An Account of English Ants,* by the Rev. W. Gould, 1747 p. 36.

perhaps a little sooner. The female eggs are covered with a thin black membrane, are oblong, and about the sixteenth or seventeenth part of an inch in length. The male eggs are of a more brown complexion, and usually laid in March.'

These dark eggs are not those of ants, but of aphides. The error is very pardonable, because the ants treat these eggs exactly as if they were their own, guarding and tending them with the utmost care. I first met with them in February 1876, and was much astonished, not being at that time aware of Huber's observations. I found, as Huber had done before me, that the ants took great care of these brown bodies, carrying them off to the lower chambers with the utmost haste when the nest was disturbed. I brought some home with me and put them near one of my own nests, when the ants carried them inside. That year I was unable to carry my observations further. In 1877 I again procured some of the same eggs, and offered them to my ants, who carried them into the nest, and in the course of March I had the satisfaction of seeing them hatch into young aphides. M. Huber, however, did not think that these were ordinary eggs. On the contrary, he agreed with Bonnet, 'that the insect, in a state nearly perfect, quits the body of its mother in that covering which shelters it from the cold in winter, and that it is not, as other germs are, in the egg surrounded by food by means of which it is developed and supported. It is nothing more than an asylum of which the aphides born at another season have no need; it is on this account some are produced naked, others enveloped in a covering. The mothers are not, then, truly oviparous, since their young are almost as perfect as they ever will be, in the asylum in which Nature has placed them at their birth.'[1]

This is, I think, a mistake. I do not propose here to describe the anatomy of the aphis; but I may observe that I have examined the female, and find these eggs to arise in the manner described by Huxley,[2] and which I have also myself observed in other aphides and in allied genera.[3] Moreover, I have opened the eggs themselves, and have also examined sections, and have satisfied myself that they are really eggs containing ordinary yelk. So far from the young insect being 'nearly perfect,' and merely enveloped in a protective membrane, no limbs or internal organs are present. In fact, the young aphis does not develop in them until shortly before they are hatched.[4]

When my eggs hatched I naturally thought that the aphides belonged to one of the species usually found on the roots of plants in the nests of *Lasius flavus*. To my surprise, however, the young creatures

[1] *The Natural History of Ants*, by M. P. Huber, 1820, p. 246.
[2] *Linnean Transactions*, 1858.
[3] *Philosophical Transactions*, 1859.
[4] I do not enter here into the technical question of the difference between ova and pseudova. I believe these to be true ova, but the point is that they are not a mere envelope containing a young aphis, but eggs in the ordinary sense, the contents of which consist of yelk, and in which the young aphis is gradually developed.

made the best of their way out of the nest, and, indeed, were sometimes brought out by the ants themselves. In vain I tried them with roots of grass &c.; they wandered uneasily about, and eventually died. Moreover, they did not in any way resemble the subterranean species. In 1878 I again attempted to rear these young aphides; but though I hatched a great many eggs, I did not succeed. In 1879, however, I was more fortunate. The eggs commenced to hatch the first week in March. Near one of my nests of *Lasius flavus*, in which I had placed some of the eggs in question, was a glass containing living specimens of several species of plant commonly found on or around ants' nests. To this some of the young aphides were brought by the ants. Shortly afterwards I observed on a plant of daisy, in the axils of the leaves, some small aphides, very much resembling those from my nest, though we had not actually traced them continuously. They seemed thriving, and remained stationary on the daisy. Moreover, whether they had sprung from the black eggs or not, the ants evidently valued them, for they built up a wall of earth round and over them. So things remained throughout the summer; but on the 9th October I found that the aphides had laid some eggs exactly resembling those found in the ants' nests; and on examining daisy-plants from outside, I found on many of them similar aphides, and more or less of the same eggs.

I confess these observations surprised me very much.

The statements of Huber, though confirmed by Schmarda, have not, indeed, attracted so much notice as many of the other interesting facts which they have recorded; because if aphides are kept by ants in their nests, it seems only natural that their eggs should also occur. The above case, however, is much more remarkable. Here are aphides, not living in the ants' nests, but outside, on the leaf-stalks of plants. The eggs are laid early in October on the food-plant of the insect. They are of no direct use to the ants, yet they are not left where they are laid, exposed to the severity of the weather and to innumerable dangers, but brought into their nests by the ants, and tended by them with the utmost care through the long winter months until the following March, when the young ones are brought out and again placed on the young shoots of the daisy. This seems to me a most remarkable case of prudence. Our ants may not perhaps lay up food for the winter; but they do more, for they keep during six months the eggs which will enable them to procure food during the following summer, a case of prudence unexampled in the animal kingdom.

The nests of our common yellow ant (*Lasius flavus*) contain in abundance four or five species of aphis, more than one of which appears to be as yet undescribed. In addition, however, to the insects belonging to this family, there are a large number of others which live habitually in ants' nests, so that we may truly say that our English ants possess a much greater variety of domestic animals than we do ourselves. Märkel satisfied himself that large nests of *Formica rufa* might contain at least a thousand of such guests;[1] and I believe that the aphides in a large nest of *Lasius flavus* would often be even more numerous. André[2] gives a list of no less than 584 species of insects, which are habitually found in association with ants, and of which 542 are beetles.

The association of some of these insects with ants may be purely accidental and without significance. In some of them no doubt the bond of union is merely the selection of similar places of abode; in some few others the ants are victimized by parasites of which they cannot rid themselves. There are, for instance, the parasitic mites, and the small black fly, belonging to the genus Phora, which lays her eggs on ants, and which I have already mentioned. Then there are some insects, such as the caterpillar of that beautiful beetle, the rosechafer, which find a congenial place of residence among the collection of bits of stick, &c., with which certain species of ants make their nests.

Another class of ant guests are those which reside actually in the galleries and chambers of, and with, the ants, but which the latter never touch. Of these the commonest in England are a species allied to Podura, for which I have proposed the name Beckia (Pl. V.

---

[1] *Beit. zur Kenntniss der unter Ameisen lebenden Insekten* Märkel, *Germar's Zeit. f. Ent.* 1841, p. 210.

[2] *Rev. et Mag. de Zool.* 1874, p. 205.

fig. 5). It is an active bustling little being, and I have kept hundreds, I may say thousands, in my nests. They run about in and out among the ants, keeping their antennæ in a perpetual state of vibration. Another very common species is a sort of white woodlouse (Pl. V. fig. 7), which enjoys the rather long name of *Platyarthrus Hoffmanseggii*. André only mentions Platyarthrus as living with *Formica rufa*, *Myrmica scabrinodis*, and *Leptothorax acervorum*. I have found it also with *Lasius niger*, *L. flavus*, and *F. fusca*. It runs about, and is evidently at home, among the ants. Both *Platyarthrus* and *Beckia*, from living constantly in the dark, have become blind; I say 'have become,' because their ancestors no doubt had eyes. In neither of these cases have I ever seen an ant take the slightest notice of either of these insects. One might almost imagine they had the cap of invisibility.

It is certain that the ants intentionally (if I may so say) sanction the residence of these insects in their nests. An unauthorised interloper would be at once killed. I have, therefore, ventured to suggest that these insects may, perhaps, act as scavengers.

In other cases the association is more close, and the ants take the greatest care of their guests.

It appears that many of these insects produce a secretion which serves as food for the ants. This is certainly the case, for instance, with the curious blind beetle, Claviger (Pl. V fig. 8), (so called from its club-shaped antennæ), which is quite blind,[1] and appears to be absolutely dependent upon the ants, as Müller first pointed out. It even seems to have lost the power of feeding itself; at any rate it is habitually fed by the ants, who supply it with nourishment as they do one another. Müller saw the ants caressing the beetles with their antennæ. The Clavigers have certain tufts of hairs at the base of the elytra, and Müller, whose observations have since been confirmed by subsequent entomologists, saw the ants take their tufts of hairs into their mouths and lick them, as well as the whole upper surface of the body, with apparently the greatest enjoyment. Grimm[2] has made a similar observation with reference to *Dinarda dentata*, another of these myrmecophilous beetles. He several times observed the ants licking the tuft of hairs at the end of the abdomen. Lespès[3] has confirmed this. On one occasion he saw an ant feed a Lomechusa. Several of the former were sucking a morsel of sugar. The beetle approached one of them, and tapped her several times on the head with her antennæ. The ant then opened her mandibles, and fed the Lomechusa as she would have done one of her own species. The Lomechusa crept on the sugar, but did not appear able to feed herself.

As might naturally be expected the myrmecophilous insects are not found indiscriminately in the nests of

---

[1] *Germar's Mag. de Zool.* 1818, p. 69.
[2] *Stettin. Ent. Zeit.* 1845, p. 123.
[3] *Ann. Soc. Ent.* France, 1855, p. 51.

ants, but while some associate with several species, many are confined to a few or even to one.

V. Hagens is of opinion [1] that in some of these beetles which frequent the nests of two or more species of ant, varieties have been produced. Thus he has observed that the specimens of *Thiasophila angulata* in nests of *Formica congerens* are darker than those found with *F. exsecta*. *Hetærius sesquicornis* found with *Lasius niger* and *Tapinoma erraticum* are smaller than those which occur in the nests of larger ants; and the form of *Dinarda dentata*, which is met with in nests of *F. sanguinea*, has rather wider wing-cases than the normal type.

I would by no means intend to imply that the relations between ants and the other insects which live with them are exhausted by the above suggestions. On the contrary, various other reasons may be imagined which may render the presence of these insects useful or agreeable to the ants. For instance, they may emit an odour which is pleasant to the ants. Again, Mr. Francis Galton has, I think, rendered it very probable that some of our domestic animals were kept as pets before they were made of any use. Unlikely as this may appear in some cases, for instance in the pig, we know as a fact that pigs are often kept by savages as pets. I would not put it forward as a suggestion which can be supported by any solid reasoning, but it

---

[1] *Berlin. Ent. Zeit.* 1865, p. 108.

seems not altogether impossible that some of these tame insects may be kept as pets.

It is from this point of view a very interesting fact that, according to Forel, in the cases of *Chennium* and *Batrisus* there is rarely more than one beetle in each nest.[1]

I now come to the relations existing between the different species of ants.

It is hardly necessary to say that, as a general rule, each species lives by itself. There are, however, some interesting exceptions. The little *Stenamma Westwoodii* (Pl. III. fig. 3) is found exclusively in the nests of the much larger *F. rufa* and the allied *F. pratensis*. We do not know what the relations between the two species are. The *Stenammas*, however, follow the *Formicas* when they change their nest, running about among them and between their legs, tapping them inquisitively with their antennæ, and even sometimes climbing on to their backs, as if for a ride, while the large ants seem to take little notice of them. They almost seem to be the dogs, or perhaps the cats, of the ants. Another small species, *Solenopsis fugax* (Pl. III. fig. 4), which makes its chambers and galleries in the walls of the nests of larger species, is the bitter enemy of its hosts. The latter cannot get at them, because they are too large to enter the galleries. The little *Solenopsis*, therefore, are quite safe, and, as it appears, make incursions into the nurseries of the larger ant, and carry off the

---

[1] *Fourmis de la Suisse*, p. 426.

larvæ as food. It is as if we had small dwarfs, about eighteen inches to two feet long, harbouring in the walls of our houses, and every now and then carrying off some of our children into their horrid dens.

Most ants, indeed, will carry off the larvæ and pupæ of others if they get a chance; and this explains, or at any rate throws some light upon, that most remarkable phenomenon, the existence of slavery among ants. If you place a number of larvæ and pupæ in front of a nest of the Horse ant (*F. rufa*), for instance, they are soon carried off; and those which are not immediately required for food remain alive for some time, and are even fed by their captors.

Both the Horse ant (*Formica rufa*, Pl. II. fig. 5) and the slave ant (*F. fusca*, Pl. I. fig. 3) are abundant species, and it must not unfrequently occur that the former, being pressed for food, attack the latter and carry off some of their larvæ and pupæ. Under these circumstances it no doubt occasionally happens that the pupæ come to maturity in the nests of the Horse ant, and it is said that nests are sometimes, though rarely, found in which, with the legitimate owners, there are a few *F. fuscas*. With the Horse ant this is, however, a very rare and exceptional phenomenon; but with an allied species, *F. sanguinea* (Pl. I. fig. 6), a species which exists in some of our southern counties and throughout Europe, it has become an established habit. The *F. sanguineas* make **periodical** expeditions, attack neighbouring nests, and carry off the pupæ. When the latter come to maturity they find themselves in a nest consisting partly of *F. sanguineas*, partly of their own species, the results of previous expeditions. They adapt themselves to circumstances, assist in the ordinary household duties, and, having no young of their own species, feed and tend those of the *F. sanguineas*. But though the *F. sanguineas* are thus aided by their slaves, or as they should rather perhaps be called, their auxiliaries, they have not themselves lost the instinct of working. It seems not improbable that there is some division of functions between the two species, but we have as yet no distinct knowledge on this point; and at any rate the *F. sanguineas* can 'do' for themselves, and carry on a nest, if necessary, without slaves.

The species usually enslaved by *F. sanguinea* are *Formica fusca* and *F. rufibarbis*, which indeed are so similar that they are perhaps varieties rather than species. Sometimes both occur in the same nest. André says that they also make slaves of *Formica gagates*.[1] Schenk asserts[2] the same of *Lasius alienus*, and F. Smith of *L. flavus*, but Forel denies these statements.[3]

Another species, *Polyergus rufescens*, is much more dependent on its slaves, being, indeed, almost entirely so.

Rev. et Mag. de Zool. 1874, p. 164.
[2] *Cat. of Brit. Foss. Hymen.*, p. 7
[3] *Fourmis de la Suisse*, p. 363.

For the knowledge of the existence of slavery among ants we are indebted to Huber,[1] and I cannot resist quoting the passage in which he records his discovery:—'On June 17, 1804,' he says, 'while walking in the environs of Geneva, between four and five in the evening, I observed close at my feet, traversing the road, a legion of Rufescent ants.

'They moved in a body with considerable rapidity and occupied a space of from eight to ten inches in length, by three or four in breadth. In a few minutes they quitted the road, passed a thick hedge, and entered a pasture ground, where I followed them. They wound along the grass without straggling, and their column remained unbroken, notwithstanding the obstacles they had to surmount. At length they approached a nest, inhabited by dark ash-coloured ants, the dome of which rose above the grass, at a distance of twenty feet from the hedge. Some of its inhabitants were guarding the entrance; but, on the discovery of an approaching army, darted forth upon the advanced guard. The alarm spread at the same moment in the interior, and their companions came forth in numbers from their underground residence. The Rufescent ants, the bulk of whose army lay only at the distance of two paces, quickened their march to arrive at the foot of the ant-hill; the whole battalion, in an instant, fell upon and overthrew the ash-coloured ants, who, after a short but obstinate conflict, retired to the bottom of

[1] *The Natural History of Ants*, by M. P. Huber, p. 249.

their nest. The Rufescent ants now ascended the hillock, collected in crowds on the summit, and took possession of the principal avenues, leaving some of their companions to work an opening in the side of the ant-hill with their teeth. Success crowned their enterprise, and by the newly-made breach the remainder of the army entered. Their sojourn was, however, of short duration, for in three or four minutes they returned by the same apertures which gave them entrance, each bearing off in its mouth a larva or a pupa.'

The expeditions generally start in the afternoon, and are from 100 to 2,000 strong.

Polyergus rufescens present a striking lesson of the degrading tendency of slavery, for these ants have become entirely dependent on their slaves. Even their bodily structure has undergone a change: the mandibles have lost their teeth, and have become mere nippers, deadly weapons indeed, but useless except in war. They have lost the greater part of their instincts: their art, that is, the power of building; their domestic habits, for they show no care for their own young, all this being done by the slaves; their industry—they take no part in providing the daily supplies; if the colony changes the situation of its nest, the masters are all carried by the slaves on their backs to the new one; nay, they have even lost the habit of feeding. Huber placed thirty of them with some larvæ and pupæ and a supply of honey in a box. 'At first,' he says, they appeared to pay some little attention to the larvæ; they carried them here and there, but presently replaced them. More than one-half of the Amazons died of hunger in less than two days. They had not even traced out a dwelling, and the few ants still in existence were languid and without strength. I commiserated their condition, and gave them one of their black companions. This individual, unassisted, established order, formed a chamber in the earth, gathered together the larvæ, extricated several young ants that were ready to quit the condition of pupæ, and preserved the life of the remaining Amazons.'

This observation has been fully confirmed by other naturalists. However small the prison, however large the quantity of food, these stupid creatures will starve in the midst of plenty rather than feed themselves.

M. Forel was kind enough to send me a nest of *Polyergus*, and I kept it under observation for more than four years. My specimens of *Polyergus* certainly never fed themselves, and when the community changed its nest, which they did several times, the mistresses were carried from the one to the other by the slaves. I was even able to observe one of their marauding expeditions, in which, however, the slaves took a part.

I do not doubt that, as Huber tells us, specimens of *Polyergus* if kept by themselves in a box would soon die of starvation, even if supplied with food. I have, however, kept isolated specimens for three months by giving them a slave for an hour or two a day to clean and feed them: under these circumstances they remained in perfect health, while, but for the slaves, they would have perished in two or three days. Excepting the slave-making ants, and some of the *Myrmecophilous* beetles above described, I know no case in nature of an animal having lost the instinct of feeding.

In *P. rufescens*, the so-called workers, though thus helpless and idle, are numerous, energetic, and in some respects even brilliant. In another slave-making ant, *Strongylognathus*, the workers are much less numerous, and so weak that it is an unsolved problem how they contrive to make slaves. In the genus *Strongylognathus* there are two species, *S. huberi* and *S. testaceus*. *S. huberi*, which was discovered by Forel, very much resembles *Polyergus rufescens* in habits. They have sabre-like mandibles, like those of *Polyergus*, and their mode of fighting is similar, but they are much weaker insects; they make slaves of *Tetramorium cæspitum*, which they carry off as pupæ. In attacking the *Tetramoriums* they seize them by the head with their jaws, just in the same way as *Polyergus*, but have not strength enough to pierce them as the latter do. Nevertheless, the *Tetramoriums* seem much afraid of them.

The other species, *Strongylognathus testaceus*, is even weaker than *S. huberi*, and their mode of life is still in many respects an enigma. They also keep the workers of *Tetramorium* in, so to say, a state of slavery, but how they procure the slaves is still a mystery. They fight in the same manner as *Polyergus*; but yet Schenk, Von Hagens, and Forel all agree that they are no match for the *Tetramoriums*, a courageous species, and one which lives in large communities. On one occasion Forel brought a nest of *Tetramorium* and put it down very near one of *Strongylognathus testaceus* with *Tetramorium* slaves. A battle at once commenced between the two communities. The *Strongylognathus* rushed boldly to the fight, but, though their side won the day, this was mainly due to the slaves. The *Strongylognathus* themselves were almost all killed; and though the energy of their attack seemed at first to disconcert their opponents, Forel assures us that they did not succeed in killing even a single *Tetramorium*. In fact, as Forel graphically observes, *Strongylognathus* is 'une triste caricature' of *Polyergus*, and it seems almost impossible that by themselves they could successfully attack a nest of *Tetramorium*. Moreover, in *Strongylognathus*, the workers are comparatively few. Nevertheless, they are always found with the *Tetramoriums*, and in these mixed nests there are no males or females of *Tetramorium*, but only those of *Strongylognathus*. Again, the whole work of the nest is done by the slaves, though *Strongylognathus* has not, like *Polyergus*, entirely lost the power of feeding itself.

But if the economy of *Strongylognathus* is an enigma, that of *Anergates* is still more mysterious.

ANERGATES.

The genus *Anergates* was discovered by **Schenk**,[1] who found a small community consisting of males, females and workers, which he naturally supposed to belong to one species. Mayr, however, pointed out[2] that the workers were in fact workers of *Tetramorium cæspitum*; and it would appear that while in *Strongylognathus* the workers are comparatively few, *Anergates* differs from all other ants in having no workers at all. The males and females live with *Tetramorium cæspitum*, and are in several respects very peculiar,— for instance, the male is wingless. One might consider it rather a case of parasitism than of slavery, but the difficulty is that in these mixed nests there are no males, females, or young of *Tetramorium*. As to this all observers are agreed. It seems quite clear that *Anergates* cannot procure its slaves, if such they are, by marauding expeditions like those of *Polyergus*; in the first place, because the *Anérgates* are too few, and secondly, because they are too weak. The whole question is rendered still more difficult by the fact that neither Von Hagens[3] nor Forel ever found either larvæ or pupæ of *Tetramorium* in the mixed nests. The community consisted of males and females of *Anergates*, accompanied and tended by workers of *Tetramorium cæspitum*. The *Anergates* are absolutely dependent

---

[1] 'Die Nassauischen Ameisen-Species,' *Stettin Ent. Zeit.* 1853, p. 186.

[2] *Europ. Formicidæ*, p. 56.

[3] *Verh. des Natur. Vereines der Preuss. Rheinlande und Westphalens* 1867, p. 53. See also V. Hagens. *Berl. Ent. Zeit.* 1867, p. 102.

upon their slaves, and cannot even feed themselves. The whole problem is, therefore, most puzzling and interesting.[1]

As regards *Strongylognathus*, Von Hagens made two suggestions, the first being that this insect is really a monstrous form of *Tetramorium*. This, however, cannot at any rate be the case with *Anergates*. On the whole, then, he inclines to think that perhaps the nests containing *Strongylognathus* or *Anergates* are only parts of a community, and that the young of the *Tetramoriums* are in another nest of the same community. This would account for the absence of the young of the *Tetramoriums*, but would not remove all the difficulties. It is in other respects not consistent with what we know of the habits of ants, and on the whole I agree with Forel in thinking the suggestion untenable.

The difficulty of accounting for the numbers of *Tetramoriums*, coupled with the absence of young, was indeed almost insuperable as long as the workers were supposed to live only for one year. My observations, however, which show that even in captivity a nest may continue for five years, place the question in a different position, and give us, I think, a clue.

On the whole, I would venture to suggest that the male and female *Anergates* make their way into a nest

[1] On the contrary, in *Tomognathus sublævis*, a Finland species which lives in the nests of *Leptothorax muscorum* and *L. acervorum*, the workers only are known. The male, like that of *Anergates*, is wingless.

of *Tetramorium,* and in some manner contrive to assassinate their queen. I have shown that a nest of ants may continue, even in captivity, for five years, without a queen. If, therefore, the female of *Anergates* could by violence or poison destroy the queen of the *Tetramoriums,* we should in the following year have a community composed of the two *Anergates,* their young, and workers of *Tetramorium,* in the manner described by Van Hagens and Forel. This would naturally not have suggested itself to them, because if the life of an ant had, as was formerly supposed, been confined to a single season, it would of course have been out of the question; but as we now know that the life of ants is so much more prolonged than had been supposed, it is at least not an impossibility.

It is conceivable that the *Tetramoriums* may have gradually become harder and stronger; the marauding expeditions would then be less fruitful and more dangerous, and might become less and less frequent. If, then, we suppose that the females found it possible to establish themselves in nests of *Tetramorium,* the present state of things would almost inevitably be, by degrees, established. Thus we may explain the remarkable condition of *Strongylognathus,* armed with weapons which it is too weak to use, and endowed with instincts which it cannot exercise.

At any rate, these four genera offer us every gradation from lawless violence to contemptible parasitism. *Formica sanguinea,* which may be assumed to have comparatively recently taken to slave-making, has not as yet been materially affected.

Polyergus, on the contrary, already illustrates the lowering tendency of slavery. They have lost their knowledge of art, their natural affection for their young, and even their instinct of feeding! They are, however, bold and powerful marauders.

In *Strongylognathus*, the enervating influence of slavery has gone further, and told even on the bodily strength. They are no longer able to capture their slaves in fair and open warfare. Still they retain a semblance of authority, and, when roused, will fight bravely, though in vain.

In *Anergates*, finally, we come to the last scene of this sad history. We may safely conclude that in distant times their ancestors lived, as so many ants do now, partly by hunting, partly on honey; that by degrees they became bold marauders, and gradually took to keeping slaves; that for a time they maintained their strength and agility, though losing by degrees their real independence, their arts, and even many of their instincts; that gradually even their bodily force dwindled away under the enervating influence to which they had subjected themselves, until they sank to their present degraded condition—weak in body and mind, few in numbers, and apparently nearly extinct, the miserable representatives of far superior ancestors, maintaining a precarious existence as contemptible parasites of their former slaves.

M. Lespès has given a short but **interesting** account of some experiments made by him on the relations existing between ants and their domestic animals, from which it might be inferred that even within the limits of a single species some communities are more advanced than others. He states that specimens of the curious blind beetle *Claviger*, which always occurs with ants, when transferred from a nest of *Lasius niger* to another which kept none of these domestic beetles, were invariably attacked and eaten. From this he infers that the intelligence necessary to keep Clavigers is not coextensive with the species, but belongs only to certain communities and races, which, so to say, are more advanced in civilisation than the rest of the species.

With reference to the statements of Lespès, I have more than once transferred specimens of *Platyarthrus* from one nest to another, and always found them received amicably. I even placed specimens from a nest of *Lasius flavus* in one of *Formica fusca* with the same result. I brought from the South of France some specimens of a different species, as yet undescribed, and put them in a nest of *Formica fusca*, where they lived for some time, and brought up more than one brood of young. These creatures, however, occur in most ants' nests, while Clavigers are only found in some.

But whether there are differences in advancement within the limits of the **same species** or not, there are certainly considerable differences between the different species, and one may almost fancy that we can trace stages corresponding to the principal steps in the history of human development.

I do not now refer to slave-making ants, which represent an abnormal, or perhaps only a temporary state of things, for slavery seems to tend in ants as in men to the degradation of those by whom it is adopted, and it is not impossible that the slave-making species will eventually find themselves unable to compete with those which are more self-dependent, and have reached a higher phase of civilisation. But putting these slave-making ants on one side, we find in the different species of ants different conditions of life, curiously answering to the earlier stages of human progress. For instance, some species, such as *Formica fusca*, live principally on the produce of the chase; for though they feed partly on the honey-dew of aphides, they have not domesticated these insects. These ants probably retain the habits once common to all ants. They resemble the lower races of men, who subsist mainly by hunting. Like them they frequent woods and wilds, live in comparatively small communities, and the instincts of collective action are but little developed among them. They hunt singly, and their battles are single combats, like those of the Homeric heroes. Such species as *Lasius flavus* represent a distinctly higher type of social life; they show more skill in architecture, may literally be said to have domesticated certain species of aphides, and may be compared to the pastoral stage of human progress—to the races which live on the produce of their flocks and herds. Their communities are more numerous; they act much more in concert; their battles are not mere single combats, but they know how to act in combination. I am disposed to hazard the conjecture that they will gradually exterminate the mere hunting species, just as savages disappear before more advanced races. Lastly, the agricultural nations may be compared with the harvesting ants.

Thus there seem to be three principal types, offering a curious analogy to the three great phases—the hunting, pastoral, and agricultural stages—in the history of human development.

CHAPTER V.

BEHAVIOUR TO RELATIONS.

Mr. Grote, in his 'Fragments on Ethical Subjects,' regards it as an evident necessity that no society can exist without the sentiment of morality. 'Everyone,' he says, 'who has either spoken or written on the subject, has agreed in considering this sentiment as absolutely indispensable to the very existence of society. Without the diffusion of a certain measure of this feeling throughout all the members of the social union, the caprices, the desires, and the passions of each separate individual would render the maintenance of any established communion impossible. Positive morality, under some form or other, has existed in every society of which the world has ever had experience.'

If this be so, the question naturally arises whether ants also are moral and accountable beings. They have their desires, their passions, even their caprices. The young are absolutely helpless. Their communities are sometimes so numerous, that perhaps London and Pekin are almost the only human cities which can compare with them. Moreover, their nests are no mere collections of independent individuals, nor even temporary associations like the flocks of migratory birds; but organised communities labouring with the utmost harmony for the common good. The remarkable analogies which, in so many ways, they present to our human societies, render them peculiarly interesting to us, and one cannot but long to know more of their character, how the world appears to them, and to what extent they are conscious and reasonable beings.

For my own part I cannot make use of Mr. Grote's argument, because I have elsewhere attempted to show that, even as regards man, the case is not by any means clear. But however this may be, various observers have recorded in the case of ants instances of attachment and affection.

Forel lays it down as a general rule that if ants are slightly injured, or rather unwell, their companions take care of them: on the other hand, if they are badly wounded or seriously ill, they are carried away from the nest, and left to perish.

Latreille, also, makes the following statement:—
'Le sens de l'odorat,' he says,[1] 'se manifestant d'une manière aussi sensible, je voulois profiter de cette remarque pour en découvrir le siége. On a soupçonné depuis longtemps qu'il résidoit dans les antennes. Je les arrachai à plusieurs fourmis fauves ouvrières, auprès du nid desquelles je me trouvois. Je vis aussitôt ces petits animaux que j'avois ainsi mutilés

[1] *Hist. Nat. des Fourmis*, p. 41.

tomber dans un état d'ivresse ou une espèce de folie. Ils erroient çà et là, et ne reconnoissoient plus leur chemin. Ils m'occupoient ; mais je n'étais pas le seul. Quelques autres fourmis s'approchèrent de ces pauvres affligées, portèrent leur langue sur leurs blessures, et y laissèrent tomber une goutte de liqueur. Cet acte de sensibilité se renouvela plusieurs fois ; je l'observois avec une loupe. Animaux compatissans ! quelle leçon ne donnez-vous pas aux hommes.'

'Jamais,' says M. de Saint Fargeau,[1] ' une Fourmi n'en rencontre une de son espèce blessée, sans l'enlever et la transporter à la fourmilière. L'y soigne-t-elle ? Je ne sais, mais je vois dans ce fait une bienveillance que je ne retrouve dans aucun autre insecte, même social.'

I have not felt disposed to repeat M. Latreille's experiment, and M. de St. Fargeau's statement is I think by no means correct ; indeed, many of my experiences seem to show not only a difference of character in the different species of ants, but that even within the limits of the same species there are individual differences between ants, just as between men.

I will commence with the less favourable aspect.

On one occasion (August 13) a worker of *Lasius niger*, belonging to one of my nests, had got severely wounded, but not so much so that she could not feed ; for though she had lost five of her tarsi, finding herself near some syrup, she crept to it and began to feed. I

[1] *Hist. Nat. des Ins. Hyménoptères*, vol. i. p. 99.

laid her gently on her back close to the entrance into the nest. Soon an ant came up to the poor sufferer, crossed antennæ with her for a moment, then went quietly on to the syrup and began to feed. Afterwards three other ants did the same; but none took any more notice of her.

August 15.—I found at 1 P.M. a *Myrmica ruginodis* which, probably in a fight with another ant, had lost the terminal portion of both her antennæ. She seemed to have lost her wits. I put her into her nest; but the others took no notice of her; and after wandering about a little, she retired into a solitary place, where she remained from 3 P.M. to 8 without moving. The following morning I looked for her at 5.30, and found her still at the same spot. She remained there till 9, when she came out. She remained out all day; and the following morning I found her dead.

Indeed, I have often been surprised that in certain cases ants render one another so little assistance. The tenacity with which they retain their hold on an enemy they have once seized is well known. M. Mocquerys even assures us that the Indians of Brazil made use of this quality in the case of wounds; causing an ant to bite the two lips of the cut and thus bring them together, after which they snip off the ant's head, which thus holds the lips together. He asserts that he has often seen natives with wounds in course of healing with the assistance of seven or eight ants' heads![1]

[1] *Ann. Soc. Ent. France,* 2 sér. tom. ii. p. 67.

Now I have often observed that some of my ants had the heads of others hanging on to their legs for a considerable time; and as this must certainly be very inconvenient, it seems remarkable that their friends should not relieve them of such an awkward encumbrance.

The behaviour of ants to one another differs also much according to circumstances; whether, for instance, they are alone, or supported by friends. An ant which would run away in the first case will defend herself bravely in the second.

If an ant is fighting with one of another species, her friends rarely come to her assistance. They seem generally (unless a regular battle is taking place) to take no interest in the matter, and do not even stop to look on. Some species, indeed, in such cases never appear to help one another; and even when the reverse is the case, as for instance in the genus *Lasius*, the truth seems to be that several of them attack the same enemy—their object being to destroy the foe, rather than to save their friend.

On one occasion several specimens of *Formica fusca* belonging to one of my nests were feeding on some honey spread on a slip of glass (May 22). One of them had got thoroughly entangled in it. I took her and put her down just in front of another specimen belonging to the same nest, and close by I placed a drop of honey. The ant devoted herself to the honey and entirely neglected her friend, whom she left to perish.

Again, some specimens of *Cremastogaster scutellaris* were feeding quietly (May 22) on some honey spread on a slip of glass, and one of them had got thoroughly mixed in it. I took her out and put her on the glass close by She could not disentangle herself; not one of her friends took the least notice of her, and eventually she died. I then chloroformed one, and put her on the board among her friends. Several touched her, but from 12 to 2.30 P.M. none took any particular notice of her.

On the other hand, I have only on one occasion seen a living ant expelled from her nest. This happened in a community of *F. fusca*. I observed (April 23, 1880) an ant carrying another belonging to the same community away from the nest. The condemned ant made a very feeble resistance. The first ant carried her burthen hither and thither for some time, evidently trying to get away from the nest, which was enclosed in the usual manner by a fur barrier. After watching for some time I provided the ant with a paper bridge, up which she immediately went, dropped her victim on the far side, and returned home. Could this have been a case in which an aged or invalid ant was being expelled from the nest?

I have often had ants in my nests to which mites had attached themselves.

Thus, on October 14, 1876, I observed that one of my ants (*Formica fusca*) had a mite attached to the underside of her head, which it almost equalled in size. The poor ant could not remove it herself, and, being a queen, never left the nest, so that I had no opportunity of doing so. For more than three months none of her companions performed this kind office for her.

With reference to this part of the subject, also, I have made some experiments.

January 3, 1876.—I immersed an ant (*Lasius niger*) in water for half an hour; and when she was then to all appearance drowned, I put her on a strip of paper leading from one of my nests to some food. The strip was half an inch wide; and one of my marked ants belonging to the same nest was passing continually to and fro over it to some food. The immersed ant lay there an hour before she recovered herself; and during this time the marked ant passed by eighteen times without taking the slightest notice of her.

I then immersed another ant in the water for an hour, after which I placed her on the strip of paper as in the preceding case. She was three-quarters of an hour before she recovered: during this time two marked ants were passing to and fro; one of them went by eighteen times, and the other twenty times; and two other ants also went over the paper; but none of them took the slightest notice of their drowned friend.

I then immersed another ant for an hour, and put her on the strip of paper. She took an hour to recover. The same two marked ants as in the previous observation were at work. One passed thirty times, the other twenty-eight times, besides which five others passed by; but not one took the slightest notice.

I immersed three ants for eight hours, and then put them on the strip of paper. They began to recover in three-quarters of an hour, but were not quite themselves till half an hour afterwards. During the first three-quarters of an hour two marked ants passed, each four times; and two others also went by. During the following half-hour the two marked ants passed sixteen times, and three others; but none of them took any notice.

I immersed another ant for forty minutes, and put her on the strip of paper. She recovered in twenty minutes, during which time the marked ones, which were the same as in the preceding case, went by fourteen times without taking any notice.

I immersed two ants for ten hours, and then placed them on the strip of paper. The same two marked ants passed respectively eighteen and twenty-six times, and one other passed by also without taking any notice. After this I left off watching.

I immersed two ants for four hours, and then put them on the strip of paper. They began to recover in an hour, during which two marked ants, not the same as in the preceding case, passed respectively twenty-eight and ten times, and two others went by; but none of them took any notice

I immersed an ant for an hour, and then put her on the same strip of paper as in the previous cases. A marked ant passed her twelve times; three others also went by but took no notice of her; but, on the other hand, a fourth picked her up and carried her off into the nest.

Again, I immersed an ant for an hour, and put her on the strip of paper. The marked ant passed twice, after which she did not return. Soon after, another ant came by and, picking up the immersed one, carried her off to the nest.

I do not bring forward these cases as proof or even as evidence that ants are less tender to friends in distress than previous observers have stated to be the case; but they certainly show that tenderness is not invariably the rule; and, especially when taken in connexion with the following cases, they are interesting illustrations of the individual differences existing between ants—that there are Priests and Levites, and good Samaritans among them, as among men.

As evidence both of their intelligence and of their affection for their friends, it has been said by various observers that when ants have been accidentally buried they have been very soon dug out and rescued by their companions. Without for one moment doubting the facts as stated, we must remember the habit which ants have of burrowing in loose fresh soil, and especially their practice of digging out fresh galleries when their nests are disturbed.

It seemed to me, however, that it would not be difficult to test whether the excavations made by ants under the circumstances were the result of this general habit, or really due to a desire to extricate their friends.

With this view I tried the following experiments:—

(1) On August 20 I placed some honey near a nest of *Lasius niger* on a glass surrounded with water, and so arranged that in reaching it the ants passed over another glass covered with a layer of sifted earth, about one-third of an inch in thickness. I then put some ants to the honey, and by degrees a considerable number collected round it. Then at 1.30 P.M. I buried an ant from the same nest under the earth, and left her there till 5 P.M., when I uncovered her. She was none the worse, but during the whole time not one of her friends had taken the least notice of her.

(2) I arranged (September 1) some honey again in the same way. At 5 P.M. about fifty ants were at the honey, and a considerable number passing to and fro. I then buried an ant as before, taking of course one from the same nest. At 7 P.M. the number of ants at the honey had nearly doubled. At 10 P.M. they were still more numerous, and had carried off about two-thirds of the honey. At 7 A.M. the next morning the honey was all gone, two or three were still wandering about, but no notice had been taken of the prisoner, whom I then let out. In this case I allowed the honey to be finished, because I thought it might perhaps be alleged that the excitement produced by such a treasure distracted their attention, or even (on the principle of doing the greatest good to the greatest number) that

they were intelligently wise in securing a treasure of food before they rescued their comrade, who, though in confinement, was neither in pain nor danger. So far as the above ants, however, are concerned, this cannot, I think, be urged.

(3) On the 8th September I repeated the experiment, burying some ants at 4 P.M. Up to 6.3 no attempt had been made to release them. I let them out and buried some more. The next morning, at 7 A.M., the honey was all gone, some ants were still wandering about, but no notice had been taken of the captives, whom I then liberated.

(4) I then (August 21) made exactly the same experiment with *Myrmica ruginodis*, as representing the other great family of ants.

In order to test the affection of ants belonging to the same nest for one another I tried the following experiments. I took six ants from a nest of *F. fusca*, imprisoned them in a small bottle, one end of which was covered with a layer of muslin. I then put the muslin close to the door of the nest. The muslin was of open texture, the meshes, however, being sufficiently small to prevent the ants from escaping. They could not only, however, see one another, but communicate freely with their antennæ. We now watched to see whether the prisoners would be tended or fed by their friends. We could not, however, observe that the least notice was taken of them. The experiment, nevertheless, was less conclusive than could be wished, because they might have been fed at night, or at some time when we were not looking. It struck me, therefore, that it would be interesting to treat some strangers also in the same manner.

On September 2, therefore, I put two ants from one of my nests of *F. fusca* into a bottle, the end of which was tied up with muslin as described, and laid it down close to the nest. In a second bottle I put two ants from another nest of the same species. The ants which were at liberty took no notice of the bottle containing their imprisoned friends. The strangers in the other bottle, on the contrary, excited them considerably. The whole day one, two, or more ants stood sentry, as it were, over the bottle. In the evening no less than twelve were collected round it, a larger number than usually came out of the nest at any one time. The whole of the next two days, in the same way, there were more or less ants round the bottle containing the strangers; while, as far as we could see, no notice whatever was taken of the friends. On the 9th the ants had eaten through the muslin, and effected an entrance. We did not chance to be on the spot at the moment; but as I found two ants lying dead, one in the bottle and one just outside, I think there can be no doubt that the strangers were put to death. The friends throughout were quite neglected.

September 21.—I then repeated the experiment, putting three ants from another nest in a bottle as before. The same scene was repeated. The friends were neglected. On the other hand, some of the ants were always watching over the bottle containing the strangers, and biting at the muslin which protected them. The next morning at 6 A.M. I found five ants thus occupied. One had caught hold of the leg of one of the strangers, which had unwarily been allowed to protrude through the meshes of the muslin. They worked and watched, though not, as far as I could see, with any system, till 7.30 in the evening, when they effected an entrance, and immediately attacked the strangers.

September 24.—I repeated the same experiment with the same nest. Again the ants came and sat over the bottle containing the strangers, while no notice was taken of the friends.

The next morning again, when I got up, I found five ants round the bottle containing the strangers, none near the friends. As in the former case, one of the ants had seized a stranger by the leg, and was trying to drag her through the muslin. All day the ants clustered round the bottle, and bit perseveringly, though not systematically, at the muslin. The same thing happened all the following day.

These observations seemed to me sufficiently to test the behaviour of the ants belonging to this nest under these circumstances. I thought it desirable, however, to try also other communities. I selected, therefore, two other nests. One was a community of *Polyergus rufescens* with numerous slaves. Close to where the ants of this nest came to feed, I placed as before two small bottles, closed in the same way—one containing two slave ants from the nest, the other two strangers. These ants, however, behaved quite unlike the preceding, for they took no notice of either bottle, and showed no sign either of affection or hatred. One is almost tempted to surmise that the warlike spirit of these ants was broken by slavery.

The other nest which I tried, also a community of *Formica fusca*, behaved exactly like the first. They took no notice of the bottle containing the friends, but clustered round and eventually forced their way into that containing the strangers.

It seems, therefore, that in these curious insects hatred is a stronger passion than affection.

Some of those who have done me the honour of noticing my papers have assumed that I disputed altogether the kindly feelings which have been attributed to ants. I should, however, be very sorry to treat my favourites so unfairly. So far as I can observe, ants of the same nest never quarrel. I have never seen the slightest evidence of ill-temper in any of my nests: all is harmony. Nor are instances of active assistance at all rare. Indeed, I have myself witnessed various cases showing care and tenderness on their part.

In one of my nests of *Formica fusca* was an ant which had come into the world without antennæ. Never having previously met with such a case, I watched her with great interest; but she never appeared to leave the nest. At length one day I found her wandering about in an aimless sort of manner, and apparently not knowing her way at all. After a while she fell in with some specimens of *Lasius flavus*, who directly attacked her. I at once set myself to separate them; but whether owing to the wounds she had received from her enemies, or to my rough, though well-meant handling, or to both, she was evidently much wounded, and lay helplessly on the ground. After some time another *Formica fusca* from her nest came by. She examined the poor sufferer carefully, then picked her up carefully and carried her away into the nest. It would have been difficult for any one who witnessed this scene to have denied to this ant the possession of humane feelings.

Again, in one of my nests of *Formica fusca* on January 23 last (1881), I perceived a poor ant lying on her back and quite unable to move. The legs were in cramped attitudes, and the two antennæ rolled up in spirals. She was, of course, altogether unable to feed herself. After this I kept my eye on her. Several times I tried uncovering the part of the nest where she was. The other ants soon carried her into the shaded part. On March 4 the ants were all out of the nest, probably for fresh air, and had collected together in a corner of the box; they had not, however, forgotten her, but had carried her with them. I took off the glass lid of the box, and after a while they returned as usual to the nest, taking her in again. On March 5 she was still alive, but on the 15th, notwithstanding all their care, she was dead!

At the present time I have two other ants perfectly crippled in a similar manner, and quite unable to move, which have lived in two different nests, belonging also to *F. fusca*, the one for five the other for four months.

In May 1879 I gave a lecture on Ants at the Royal Institution, and was anxious to exhibit a nest of *Lasius flavus* with the queen. While preparing the nest, on May 9, we accidentally crushed the queen. The ants, however, did not desert her, or drag her out as they do dead workers, but, on the contrary, carried her with them into the new nest, and subsequently into a larger one with which I supplied them, congregating round her, just as if she had been alive, for more than six weeks, when we lost sight of her.

In order to ascertain whether ants knew their fellows by any sign or pass word, as has been suggested in the case of bees, I was anxious to see if they could recognise them when in a state of insensibility. I tried therefore the following experiments with some specimens of *Lasius flavus*.

September 10, at 6 P.M., a number of these ants were out feeding on some honey, placed on one of my tables, and surrounded by a moat of water. I chloroformed four of them and also four from a nest in my park, at some distance from the place where the first had been originally procured, painted them, and put them close to the honey. Up to 8.20 the ants had taken no notice of their insensible fellow creatures. At 9.20 I found that four friends were still lying as before, while the four strangers had been removed. Two of them I found had been thrown over the edge of the board on which the honey was placed. The other two I could not see.

Again, on September 14, at 8.40, I put in the same way four friends marked white, and four strangers marked red, close to where my *L. flavus* were out feeding on honey placed on a slip of glass over water. For some hours they took no notice of them. At length one took a friend, and after carrying her about some time dropped her, at 12.40, into the water. Some time after another took up a stranger and carried her into the nest at 2.35. A second stranger was similarly carried into the nest at 2.55, a third at 3.45, while the fourth was thrown over the edge of the board at 4.20. Shortly after this two of the strangers were brought out of the nest again and thrown into the water. A second friend was thrown away, like the first, at 4.58, the third at 5.17, and the fourth at 5.46. I could not ascertain what happened to the last stranger, but have little doubt that she was brought out of the nest and thrown away like the rest.

On the following day at 6.45 I tried the same experiment again, only reversing the colors by which they were distinguished. At 7 one of the strangers was carried off and dropped over the edge of the glass into the water, and at 8 a second. At 8.45 a friend was taken up and, after being carried about some time, was thrown into the moat. At 9.45 a friend was picked up and carried into the nest, but brought out again and thrown away about 3 in the afternoon. The other four remained where they were placed until 8 P.M., and though the other ants often came up and examined them, they did not carry them off.

September 29.—Again placed nine chloroformed ants, five friends and four strangers, close to where a number were feeding. There was a continual stream of ants to the honey, ten or fifteen being generally there at once.

A stranger was picked up at 10.20 {and dropped into the water at} 10.32

"	"	"	10.22	"	10.35
A friend	"	"	11.22	"	11.42
A stranger	"	"	11.35	"	11.50
"	"	"	11.41	"	11.45

Shortly after the others were picked up and carried away to the edge of the board, where they were dropped, but none were taken into the nest.

October 2.—Again at 10 A.M. placed ten chloroformed ants, five friends and five strangers, close to where some were feeding. They were picked up and carried off as before in the following order:—

At 11. 5 a stranger was picked up and dropped at 11.15

11.12 a friend	"	"	11.50
11.25 a stranger	"	"	11.36
12. 7 "	"	"	12.45
12.10 a friend	"	"	12.16

At 1.10 a stranger was picked up and dropped at 2. 6
   1.42 a friend   ,,   ,,   1.46
   1.52   ,,   ,,   ,,   1.56
   2.6   ,,   ,,   ,,   3.10

Only one of them, and that one a stranger, was carried into the nest at 12.45, but brought out again at 1.10.

October 6.—At 9 A.M. again tried the same experiment with four strangers and five friends.

At 9.25 a friend was picked up and dropped at 9.31
   9.32   ,,   ,,   ,,   9.38
   9.35 a stranger   ,,   ,,   9.45
   9.45 a friend   ,,   ,,   9.52
   10. 8 a stranger   ,,   ,,   10.17
   10.17 a friend   ,,   ,,   10.20
   10.22 a stranger   ,,   ,,   10.25
   10.28   ,,   ,,   ,,   10.40
   10.25 a friend   ,,   ,,   10.31

None of them were carried into the nest.

These experiments seem to prove that under such circumstances ants, at least those belonging to this species, do not carry off their friends (when thus rendered insensible) into a place of safety.

I think, however, that in this experiment the ants being to all intents and purposes dead, we could not expect that any difference would be made between friends and strangers. I therefore repeated the same experiment, only, instead of chloroforming the ants, I intoxicated them. This experiment is more difficult, as it is not in all cases easy to hit off the requisite degree of intoxication. The numbers therefore of friends and strangers are not quite the same, because in some cases the ants recovered too quickly and had to be removed. In such cases I have latterly replaced the ant so removed by another, so as to keep the number of friends and strangers about equal. The sober ants seemed somewhat puzzled at finding their intoxicated fellow creatures in such a disgraceful contion, took them up, and carried them about for a time in a somewhat aimless manner.

November 20.—I experimented with six friends and six strangers, beginning at 11.

At 11.30 a friend was carried to the nest.

11.50 a stranger was dropped into the water.

12.30 ,, ,, ,,

12.31 a friend ,, ,,

1.10 a stranger ,, ,,

1.18 ,, ,, ,,

1.27 ,, ,, ,,

1.30 a friend (partly recovered) was taken to the nest.

2.30 a friend was taken up and carried about till 2.55; she was then taken to the nest, but at the door the bearer met two other ants, which seized the intoxicated one, carried her off, and eventually dropped her into the water.

At 3.35 a friend was carried to the nest.

Out of these twelve, five strangers and two friends

were dropped into the water; none of the strangers, but three friends were taken to the nest. None of the friends were brought out of the nest again.

November 22.—Experimented in the same way on four friends and four strangers, beginning at 12 o'clock.

At 12.16 a stranger was taken and dropped into the water.

  12.21   ,,   ,,   ,,
  12.23   ,,   ,,   ,,
  12.40   ,,   ,,   ,,

I then put four more strangers treated as before.

At 3.10 a stranger was taken and dropped into the water.

  3.30   ,,   ,,   ,,
  3.35   ,,   ,,   ,,

3.44 a friend (partly recovered) was taken back to the nest.

4.10 a stranger was taken and dropped into the water.

4.13 a friend (partly recovered) was taken back to the nest.

In this case eight strangers were dropped into the water, and none were taken to the nest; two friends, on the contrary, were taken to the nest, and none were dropped into the water.

December 1.—Experimented with five friends and five strangers, beginning at 2.15.

At 2.30 a stranger was dropped into the water.

  3. 2   ,,   ,,   ,,

At 3.20 a friend was taken into the nest.
    3.35 a stranger was taken into the nest, but afterwards brought out again and thrown into the water.
    3.52   ,,   ,,   ,,   ,,
    4. 5 I put out four more friends and as many strangers.
    4.45 a stranger was dropped into the water.
    5.10   ,,   taken into the nest, but afterwards brought out and thrown into the water.
    5.24   ,,   taken into the nest, but afterwards brought out and thrown into the water.
    5.55 a friend was thrown into the water.
    6. 4. a stranger   ,,   ,,
    6. 4   ,,   ,,   ,,
    6. 8 a friend was taken into the nest.
    6.20   ,,   ,,   ,,
    6.23   ,,   ,,   ,,
    6.30 a stranger was dropped into the water.
    6.50 a friend   ,,   ,,   ,,
    8. 5 a friend was taken into the nest.

In this case two friends were thrown into the water and seven taken into the nest; while six strangers were thrown into the water and four were taken into the nest; all of these, however, were afterwards brought out again and thrown away.

    December 8.—Experimented with six friends and six strangers, beginning at 11.30

At 11.30 a friend was carried to nest.
   11.47 ,, ,,
   11.50 ,, ,,
   11.52 ,, ,,
   11.56 a friend was dropped into water.
   11.58 a stranger ,, ,,
   11.58 ,, ,, ,,
   12 a stranger was carried to nest.
   12. 2 ,, ,, ,,
   12. 3 ,, ,, ,, .

I then put four more of each, and as a friend or a stranger was carried off, replaced her by another.

   At 12.45 a friend to the water.
     12.58 a stranger to the water.
      1 a friend to the nest.
      1 ,, ,,
      1 ,, ,,
      1.58 ,, ,,
      1.59 ,, ,,
      2.30 a stranger to the water.
      2.30 ,, ,,
      2.35 a stranger to the nest.
      2.42 a stranger to the water.
      2.48 ,, ,,
      2.51 ,, ,,
      2.52 ,, ,,
      2.55 a friend to the nest.
      2·55 a stranger to the water.
      2.55 ,, ,,

At 3. 2 a friend to the water.
    3. 6 a stranger to the water.
    3.12 a friend to the water.
    3.15   ,,   ,,
    3.16 a friend to the nest.
    3.22 a stranger to the water
    3.25   ,,   ,,
    3.25 a friend to the nest.
    3.35 a stranger to the water
    3.50 a friend to the nest.
    3.50   ,,   ,,

All these ants appeared quite insensible. Altogether sixteen friends were taken to the nest and five thrown into the water, while of the strangers only three were taken to the nest, and fifteen were thrown into the water. Moreover, as in the preceding observation, even the three strangers which were at first taken to the nest were soon brought out again and thrown away; while this was not the case with any of the friends as far as we could ascertain, though we searched diligently for them also. In this case also all the intoxicated ants were motionless and apparently insensible.

January 15.—Repeated the same experiment, beginning at 12.20. Up to 7 P.M. not one of the intoxicated ants had been moved. At 8.20 we found a stranger in the water, at 9.30 another, and at the following morning a third. The others were untouched.

January 17.—Repeated the same experiment, beginning at 11.30.

At 12 a friend was carried to the nest.
    12.20 a stranger was dropped into the water.
    12.34 a friend was carried to the nest.
    12.40 a stranger was dropped into the water.
    12.45 a friend was carried to the nest.
    1   a stranger   ,,   ,,
    1   ,,   ,,   water
        (Stopped observing till 2.)
    2.30 a stranger was dropped into the water.
    2.30 a stranger was carried to the nest.
    4.10   ,,   ,,   ,,
    4.30 a friend   ,,   ,,
    6.20 a stranger   ,,   water.
    6.35   ,,   ,,   ,,

Thus, then, the general results were that the ants removed forty-one friends and fifty-two strangers. Of the friends, thirty-two were carried into the nest and nine were thrown into the water. Of the strangers, on the contrary, forty-three were thrown into the water; only nine were taken into the nest, and seven of these were shortly afterwards brought out again and thrown away. Indeed, I fully believe that the other two were treated in the same manner, though we could not satisfy ourselves of the fact. But it was only by very close observation that the seven were detected, and the other two may well have escaped notice.

It seems clear, therefore, that even in a condition of insensibility these ants were recognised by their friends.

Tabular View.—Experiments on Ants under Chloroform and Intoxicated.

	CHLOROFORMED ANTS.					
	FRIENDS			STRANGERS		
	To Nest	To Water	Unre-moved	To Nest	To Water	Unre-moved
Sept. 10	...	...	4	...	4	...
14	...	4	...	2 and brought out again	2	...
15	1 and brought out again	1	...	...	2	2
29	...	5	...	...	4	...
Oct. 2	...	5	...	1 and brought out again	4	...
6	...	5	...	...	4	...
	1	20	4	3	20	2

	INTOXICATED ANTS.					
Nov. 20	3	2	...	...	5	1
22	2	...	2	...	8	...
In these cases some of the Ants had partly recovered; in the following they were quite insensible.						
Dec. 1	7 none brought out again	2	...	3 all these brought out again	6	...
8	16 none brought out again	5	...	3 all these brought out again	15	...
Jan. 15	...	...	4	...	3	1
17	4 none brought out again	...	...	3 one brought out again	6	...
	27	7	4	2	30	1

CHAPTER VI.

RECOGNITION OF FRIENDS.

It has been already shown that with ants, as with bees, while the utmost harmony reigns between those belonging to the same community, all others are enemies. I have already given ample proof that a strange ant is never tolerated in a community. This of course implies that all the bees or ants of a community have the power of recognising one another, a most surprising fact, when we consider the shortness of their life and their immense numbers. It is calculated that in a single hive there may be as many as 50,000 bees, and in the case of ants the numbers are still greater. In the large communities of *Formica pratensis* it is probable that there may be as many as from 400,000 to 500,000 ants, and in other cases even these large numbers are exceeded.

If, however, a stranger is put among the ants of another nest, she is at once attacked. On this point I have satisfied myself, as will be seen in the following pages, that the statements of Huber and others are perfectly correct. If, for instance, I introduced a stranger into one of my nests, say of *Formica fusca* or

Lasius niger, she was at once attacked. One ant would seize her by an antenna, another by a leg, and she was either dragged out of the nest or killed.

Moreover, we have not only to deal with the fact that ants know all their comrades, but that they recognise them even after a lengthened separation.

Huber mentions that some ants which he had kept in captivity having accidentally escaped, 'met and recognised their former companions, fell to mutual caresses with their antennæ, took them up by their mandibles, and led them to their own nests; they came presently in a crowd to seek the fugitives under and about the artificial ant-hill, and even ventured to reach the bell-glass, where they effected a complete desertion by carrying away successively all the ants they found there. In a few days the ruche was depopulated. These ants had remained four months without any communication.'[1] This interesting statement has been very naturally copied by succeeding writers. See, for instance, Kirby and Spence's 'Introduction to Entomology,' vol. ii. p. 66, and Newport, 'Trans. of the Entomological Society of London,' vol. ii. p. 239.

Forel, indeed, regards the movements observed by Huber as having indicated fear and surprise rather than affection; though he is quite disposed to believe, from his own observations, that ants would recognise one another after a separation of several months.

[1] Huber, p. 172.

RECOGNITION OF FRIENDS. 121

The observation recorded by Huber was made casually, and he did not take any steps to test it by subsequent experiments. The fact, however, is of so much importance that I determined to make further observations on the subject. In the first place, I may repeat that I have satisfied myself by many experiments, that ants from one community introduced into another,—always be it understood of the same species,—are attacked, and either driven out or killed. It follows, therefore, that as within the nest the most complete harmony prevails—indeed, I have never seen a quarrel between sister ants—they must by some means recognise one another.

When we consider their immense numbers this is sufficiently surprising; but that they should recognise one another, as stated by Huber, after a separation of months, is still more astonishing.

I determined therefore to repeat and extend his observations.

Accordingly, on *August* 20, 1875, I divided a colony of *Myrmica ruginodis*, so that one half were in one nest, A, and the other half in another, B, and were kept entirely apart.

On *October* 3, I put into nest B a stranger and an old companion from nest A. They were marked with a spot of colour. One of them immediately flew at the stranger; of the friend they took no notice.

October 18.—At 10 A.M. I put in a stranger and a

friend from nest A. In the evening the former was killed, the latter was quite at home.

October 19.—I put one in a small bottle with a friend from nest A. They did not show any enmity. I then put in a stranger; and one of them immediately began to fight with her.

October 24.—I again put into the nest a stranger and a friend. The former was attacked, but not the latter. The following day I found the former almost dead, while the friend was all right.

October 31.—I again put a stranger and a friend into the nest. The former was at once attacked; in this case the friend also was, for a moment, seized by the leg, but at once released again. On the following morning the stranger was dead, the friend was all right.

November 7.—Again I put in a stranger and a friend. The former was soon attacked and eventually driven out; of the latter they did not seem to me to take any particular notice. I could see no signs of welcome, no gathering round a returned friend; but, on the other hand, she was not attacked.

Again, I separated one of my colonies of *Formica fusca* into two halves on August 4, 1875, and kept them entirely apart. From time to time I put specimens from the one half back into the other. The details of this experiment will be found in the Appendix. At first the friends were always amicably received, but after some months' separation they were occasionally attacked, as if some of the ants, perhaps the young ones, did not recognise them. Still they were never killed, or driven out of the nest, so that evidently when a mistake was made, it was soon recognised. No one who saw the different manner in which these ants and strangers were treated, could have the slightest doubt that the former were recognised as friends and the latter as enemies. The last three were put back on May 14, 1877, that is to say, after a separation of a year and nine months, and yet they were amicably received, and evidently recognised as friends!

These observations were all made on *Formica fusca*, and it is of course possible that other species would behave in a different manner.

Indeed, in this respect *Lasius flavus* offers a surprising contrast to *F. fusca*. I was anxious to see whether the colonies of this species, which are very numerous round my house, were in friendly relations with one another. With this view, I kept a nest of *L. flavus* for a day or two without food, and then gave them some honey, to which they soon found their way in numbers. I then put in the midst of them an ant of the same species from a neighbouring nest; the others did not attack, but, on the contrary, cleaned her—though, from the attention she excited and the numerous communications which took place between her and them, I am satisfied that they knew she was not one of themselves. After a few minutes she accompanied some of the returning ants to the nest. They did not drag nor apparently guide her; but she went with the rest freely. This I repeated several times with the same result.

I then took four ants, two from a nest about 500 yards from the first in one direction, the other from an equal distance in another. In all cases the result was the same. I then got a few from a colony about half a mile off. These also were most amicably received, and in every case the stranger went of her own accord to the nest. One of the strangers was, indeed, dragged about half way to the entrance of the nest, but was then left free and might have run away if she had liked. She, however, after wandering about for half a minute, voluntarily entered the nest. In one or two cases the stranger ran as quickly and straight to the nest as if she had been there over and over again. This, I suppose, can only have been by scent; and certainly no hounds in full cry could have pursued their game more directly or with less hesitation. In other cases, however, they were much longer before they went in. To satisfy myself that these facts were not owing to the nest having been taken from that of colonies or allies, I subsequently procured some ants of the same species from a nest in Hertfordshire; and they also behaved in a similar manner. In one or two cases they seemed to be attacked, though so feebly that I could not feel sure about it; but in no case were the ants killed.

The following fact surprised me still more. I put an ant (Aug. 13) at 9 A.M. on a spot where a number of

Lasius flavus (belonging to one of my nests of domesticated ants) had been feeding some hours previously, though none were there, or, indeed, out at all, at the moment. The entrance to the nest was about eight inches off; but she walked straight to it and into the nest. A second wandered about for four or five minutes, and then went in; a third, on the contrary, took a wrong direction, and, at any rate for three-quarters of an hour, did not find the entrance.

At that time, however, I did not ascertain what became of the specimens thus introduced into a strange community. I thought it would be worth while to determine this, so I subsequently (1881) took six ants from one of my nests of *L. flavus*, marked them, and introduced them into another nest of the same species. As in the preceding cases they entered quite readily; but though they were not at first attacked, they were evidently recognised as strangers. The others examined them carefully, and at length they were all driven out of the nest. Their greater readiness to enter a strange nest may perhaps be accounted for by the fact that, as a subterranean species their instinct always is to conceal themselves underground, whereas, *F. fusca*, a hunting species, does not do so except to enter its own nest.

How do these ants and bees recognise their companions? The difficulty of believing that in such populous communities every individual knows every other by sight, has led some entomologists to suppose that each nest had a sign or password. This was, for instance, the opinion of Gélieu, who believed that in each hive the bees had some common sign or password. As evidence of this, he mentions [1] that one of his hives had been for some days robbed by the bees from another: 'et je désespérais de conserver cet essaim, lorsqu'un jour, sur le soir, je le vis fort inquiet, fort agité, comme s'il eût perdu sa reine. Les abeilles couraient en tout sens sur le devant et le tablier de la ruche, se flairant, se tâtant mutuellement, comme si elles eussent voulu se dire quelque chose. C'était pour changer leur signe de reconnaissance, qu'elles changèrent en effet pendant la nuit. Toutes les pillardes qui revinrent le lendemain, furent arrêtées et tuées. Plusieurs échappèrent aux gardes vigilantes qui défendaient l'entrée avertirent sans doute les autres du danger qu'elles avaient couru, et que l'on ne pouvait plus piller impunément. Aucune de celles qui voulurent recommencer leur déprédation ne pénétra dans la ruche, dont elles avaient fait leur proie, et qui prospéra merveilleusement.'

Dujardin doubts the explanation given by Gélieu. He thinks that the nest which was robbed was at that time queenless, and that the sudden change in the behaviour of the bees was due to their having acquired a queen.

Burmeister, on the contrary, in his excellent 'Manual of Entomology,' says that 'the power of com-

[1] *Le Conservateur des Abeilles*, p. 143.

municating to their comrades what they purpose is peculiar to insects. Much has been talked of the so-called signs of recognition in bees, which is said to consist in recognising their comrades of the same hive by means of peculiar signs. This sign serves to prevent any strange bee from entering into the same hive without being immediately detected and killed. It, however, sometimes happens that several hives have the same signs, when their several members rob each other with impunity. In these cases the bees whose hive suffers most alter their signs, and then can immediately detect their enemy.'[1]

Others, again, have supposed ants recognise one another by smell.

Mr. McCook states that ants more or less soaked in water are no longer recognised by their friends, but, on the contrary, are attacked. Describing the following observation, he says :[2]—' I was accidentally set upon the track of an interesting discovery. An ant fell into a box containing water placed at the foot of a tree. She remained in the liquid several moments and crept out. Immediately she was seized in a hostile manner, first by one, then another, then by a third: the two antennæ and one leg were thus held. A fourth ant assaulted the middle thorax and petiole. The poor little bather was thus dragged helplessly to and fro for a long time, and was evidently ordained to death.

[1] Burmeister's *Entomology*, p. 502.
[2] *Mound-making Ants of the Alleghanies*, p. 280.

Presently I took up the struggling heap. Two of the assailants kept their hold; one finally dropped, the other I could not tear loose, and so put the pair back upon the tree, leaving the doomed immersionist to her hard fate.'

After recording one or two other similar observations, he adds: [1]—'The conclusion, therefore, seems warranted that the peculiar odour or condition by which the ants recognise each other was temporarily destroyed by the bath, and the individuals thus "tainted" were held to be intruders, alien and enemy. This conclusion is certainly unfavourable to the theory that any thing like an intelligent social sentiment exists among the ants. The recognition of their fellows is reduced to a mere matter of physical sensation or "smell."' This conclusion does not, I confess, seem to me to be conclusively established.

We can hardly suppose that each ant has a peculiar odour, and it seems almost equally difficult, considering the immense number of ants' nests, to suppose that each community has a separate and peculiar smell. Moreover, in a previous chapter I have recorded some experiments made with intoxicated ants. It will be remembered that my ants are allowed to range over a table surrounded by a moat of water. Now, as already mentioned, out of forty-one intoxicated friends, thirty-two were carried into the nest and nine were thrown into the water; while out of fifty-two intoxicated

[1] *Mound-making Ants of the Alleghanies*, p. 281.

strangers two were taken into the nest and fifty were thrown into the water. I think it most probable that even these two were subsequently brought out and treated like the rest.

It is clear, therefore, that in these species, and I believe in most, if not all others, the ants of a community all recognise one another. The whole question is full of difficulty. It occurred to me, however, that experiments with pupæ might throw some light on the subject. Although all the communities are deadly enemies, still if larvæ or pupæ from one nest are transferred to another, they are tended with apparently as much care as if they really belonged to the nest. In ant-warfare, though sex is no protection, the young are spared, at least when they belong to the same species. Moreover, though the habits of ants are greatly changed if they are taken away from their nest and kept with only a few friends, still, under such circumstances, they will carefully tend any young who may be confided to them. Now if the recognition were individual—if the ants knew any one of their comrades, as we know our friends, not only from strangers, but from one another —then young ants taken from the nest as pupæ and restored after they had come to maturity would not be recognised as friends. On the other hand, if the recognition were effected by means of some signal or password, then the pupæ which were intrusted to ants from another nest would have the password, if any, of that nest; and not of their own. Hence in this case they would be amicably received in the nest from which their nurses had been taken, but not in their own.

In the first place, therefore, I put, on September 2, 1877, some pupæ from one of my nests of *Formica fusca* with a couple of ants from the same nest. On the 27th I put two ants, which in the meantime had emerged from one of these pupæ, back into their own nest at 8.30 A.M., marking them with paint as usual. At 9 they seemed quite at home; at 9.30, ditto; at 10, ditto; and they were nearly cleaned. After that I could not distinguish them.

On the 29th another ant came out of the pupa-state; and on October 1 at 7.45 I put her back into the nest. She seemed quite at home, and the others soon began to clean her. We watched her from time to time, and she was not attacked; but, the colour being removed, we could not recognise her after 9.30.

On July 14 last year (1878) I put into a small glass some pupæ from another nest of *Formica fusca* with two friends.

On August 11 I put four of the young ants which had emerged from these pupæ into the nest. After the interval of an hour, I looked for them in vain. The door of the nest was closed with cotton-wool; so that they could not have come out; and if any were being attacked, I think we must have seen it. I believe, therefore, that in the meantime they had been cleaned. Still, as we did not actually watch them, I was not satisfied. I put in, therefore, two more at 5 P.M. At 5.30 they were all right; at 5.45, ditto, one being almost cleaned. At 6 one was all right; the other was no longer recognizable, having been quite cleaned. At 6.30 also one was quite at home; the other could not be distinguished. At 7 both had been completely cleaned

The following day I marked another, and put her in at 6 A.M. At 6.15 she was all right among the others, and also at 6.30, 7, 7.30, 8, and 9.30, after which I could no longer distinguish her.

Again, on the following day I put in another at 6.45 A.M. At 7 she was quite at home, and also at 7.15, 7.30, 8, and to 9.30, after which I did not watch her.

To test the mode in which the ants of this nest would behave to a stranger, I then, though feeling no doubt as to the result, introduced one. The difference was very striking. The stranger was a powerful ant; still she was evidently uncomfortable, started away from every ant she met, and ran nervously about, trying to get out of the nest. She was, however, soon attacked.

Again, on October 1 some pupæ of *Lasius niger* were placed in a glass with five ants from the same nest.

On December 8 I took three of the ants which had emerged from these pupæ, and at midday put them back into their old nest, having marked them by nicking the claws. Of course, under these circumstances we could not watch the ants. I examined the nest, however, every half hour very carefully, and am satisfied that there was no fighting. The next morning there was no dead ant; nor was there a death in the nest for more than a fortnight.

December 21.—Marked three more in the same manner, and put them in at 11.15 A.M. Looked at the usual intervals, but saw no fighting. The next morning there was no dead one outside the nest; but I subsequently found one of these ants outside, and nearly dead. I am, however, disposed to think that I had accidentally injured this ant.

December 23.—Painted three, and put them in at 10 A.M. At 11 they were all right, 12 ditto, 1 ditto, 2 ditto, 3 ditto, 4 ditto, 5 ditto. At 3 I put in three strangers for comparison: two of them were soon attacked; the other hid herself in a corner; but all three were eventually dragged out of the nest. I found no other dead ant outside the nest for some days.

December 29.—Painted three more, and put them in at 10.30 A.M. At 11 they were all right, 12 ditto, 1 ditto, 2 ditto. During the afternoon they were once or twice attacked for a minute or two, but the ants seemed soon to perceive the mistake, and let them go again. The next morning I found one dead ant, but had no reason to suppose that she was one of the above three. The following morning there was again only one dead ant outside the nest; she was the third of the strangers put in on the 23rd, as mentioned above. Up to January 23 found no other dead one.

January 3, 1879.—Painted three more, and put them in at 11.30 A.M. At 12 two were all right: we could not see the third; but no ant was being attacked. 12 ditto. 1, all three are all right; 2 ditto; 5 ditto. As already mentioned, for some days there was no dead ant brought out of the nest.

' *January* 5.—Painted three more and put them in at 11.30 A.M. At 12 two were all right among the others; I could not find the third; but no ant was being attacked. 12.30 ditto, 1 ditto, 2 ditto, 4 ditto. On the following morning I found two of them all right among the others. There was no dead ant.

January 13.—Painted three more and put them in at 12.30. At 1 they were all right. 2 ditto. 4, two were all right; I could not see the third, but she was not being attacked. The next morning, when I looked at the nest, one was just being carried, not dragged, out. The ant carried her about 6 inches and then put her down, apparently quite unhurt. She soon returned into the nest, and seemed to be quite amicably received by the rest. Another one of the three also seemed quite at home. The third I could not see; but up to January 23 no dead one was brought out of the nest.

January 19.—Marked the last three of these ants, and put them into the nest at 9.30 A.M. They were watched continuously up to 1. At that time two of them had been almost completely cleaned. One was attacked for about a minute soon after 11, and another a little later; but with these exceptions they were quite amicably received, and seemed entirely at home among the other ants.

Thus every one of these thirty-two ants was amicably received.

These experiments, then, seem to prove that ants removed from a nest in the condition of pupæ, but tended by friends, if reintroduced into the parent nest, are recognised and treated as friends. Nevertheless the recognition does not seem to have been complete. In several cases the ants were certainly attacked, though only by one or two ants, not savagely, and only for a short time. It seemed as if, though recognised as friends by the great majority, some few, more ignorant or more suspicious than the rest, had doubts on the subject, which, however, in some manner still mysterious, were ere long removed. The case in which one of these marked ants was carried out of the nest may perhaps be explained by her having been supposed to be ill, in which case, if the malady is considered to be fatal, ants are generally brought out of the nest.

It now remained to test the result when the pupæ were confided to the care of ants belonging to a different nest, though, of course, the same species.

I therefore took a number of pupæ out of some of my nests of *Formica fusca* and put them in small glasses, with ants from another nest of the same species. Now, as already mentioned, if the recognition were effected by means of some signal or password, then, as we can hardly suppose that the larvæ or pupæ would be sufficiently intelligent to appreciate, still less to remember it, the pupæ which were intrusted to ants from another nest would have the password, if any, of that nest, and not of the one from which they had been taken. Hence, if the recognition were effected by some password or sign with the antennæ, they would be amicably received in the nest from which their nurses had been taken, but not in their own.

I will indicate the nests by the numbers in my note-book.

On August 26 last year I put some pupæ of *Formica fusca* from one of my nests (No. 36) with two workers from another nest of the same species. Two emerged from the chrysalis state on the 30th; and on September 2 I put them, marked as usual, into their old nest (No. 36) at 9.30 A.M. At 9.45 they seemed quite at home, and had already been nearly cleaned. At 10.15 the same was the case, and they were scarcely distinguishable. After that I could no longer make them out; but we watched the nest closely, and I think I can undertake to say that if they had been attacked we must have seen it.

Another one of the same batch emerged on August 18, but was rather crippled in doing so. On the 21st I put her into the nest (No. 36). This ant was at once attacked, dragged out of the nest, and dropped into the surrounding moat of water.

Again, on July 14 last year (1878) I put some pupæ of *Formica fusca* from nest No. 36 into a glass with three ants of the same species from nest No. 60.

On the 22nd I put an ant from one of these pupæ into her old nest (No. 36) at 9.30 A.M. She was attacked. At 10 she was being dragged about. 10.30 ditto. I regretted she was not watched longer.

August 8.—Put another ant which had emerged from one of these pupæ into her old nest (No. 36) at 7.45 A.M. At 8 she seemed quite at home among the others. 8.15 ditto, 8.30 ditto, 9 ditto, 9.30 ditto.

August 9.—Put two other young ants of this batch into their old nest (No. 36) at 7 A.M. At 7.30 they were all right. At 7.30 one of them was being dragged by a leg, but only, I think, to bring her under shelter, and was then let go. Young ants of this species, when the nest is disturbed, are sometimes dragged to a place of safety in this way. At 8.30 they were all right and nearly cleaned. After this I could not distinguish them; but if they had been attacked, we must have seen it.

August 11.—Put in another one as before at 8.30 A.M. At 8.45 she was all right. At 9 she was dragged by a leg, like the last, but not for long; and at 9.30 she was quite comfortable amongst the others. 10 ditto, 10.45 ditto, 12 ditto, 5 ditto.

August 24.—Put in the last two ants of this lot

as before at 9.15 A.M. At 9.30 they were all right. 9.45 ditto. At 10 they were almost cleaned. At 10.30 I could only distinguish one; and she had only a speck of colour left. She appeared quite at home; and though I could no longer distinguish the other, I must have seen it if she had been attacked.

Thus, then, out of seven ants of this batch put back into their old nest, six were amicably received. On the other hand, I put one into nest No. 60, from which the three nurses were taken. She was introduced into the nest at 8.15 A.M., and was at once attacked. 8.45, she was being dragged about. 9 ditto, 9.15 ditto, 9.30 ditto. Evidently therefore she was not treated as a friend.

Again, on July 14, 1878, I put some pupæ of *Formica fusca* from nest No. 60 with three ants from nest No. 36.

On August 5 at 4 P.M. I put an ant which had emerged from one of these pupæ, into her old nest (No. 60). At 4.15 she seemed quite at home. They were already cleaning her; and by 4.30 she was no longer distinguishable. We watched the nest, however, carefully for some time; and I feel sure she was not attacked.

August 6.—Put another of this batch into nest No. 60 at 7.15 A.M. At 7.30 she was not attacked. At 8, one of the ants was carefully cleaning her. At 8.15 she was quite at home among the others. At 8.30 ditto; she was nearly cleaned. 9.30 ditto.

August 8.—Put in another as before at 7.45. At 8 she was all right. 8.30 ditto, 9.30 ditto, 9.45 ditto.

August 9.—Put in another as before at 7 A.M. At 7.30 she is quite at home among the others, and already nearly cleaned. At 8 I could no longer distinguish her; but certainly no ant was being attacked. 9 ditto.

August 11.—Put in another as before at 8 A.M. At 8.15 she was quite at home. 8.30 ditto, 9 ditto, 9.30 ditto, 10 ditto, 12.30 ditto.

August 13.—Lastly, I put in the remaining young ant as before at 7 A.M. At 7.15 she was all right. At 7.30 ditto and nearly cleaned. At 8 I could no longer distinguish her; but no ant was being attacked.

Thus, then, as in the preceding experiment, these six ants when reintroduced into the nest from which they had been taken as pupæ, were received as friends. On the other hand, on August 5 I put a young ant of the same batch into nest No. 36, from which the three nurses had been taken. She was introduced at 11 and was at once attacked. At 11.30 she was being dragged about, and shortly after was dragged out of the nest. I then introduced a second; but she was at once attacked like the first.

August 22.—I put some pupæ of *Formica fusca* from nest No. 64 under the charge of three ants from nest No. 60. By September 7 several young ones had emerged. I put two of them into nest No. 64 at 8.15 A.M. They were amicably received, as in the preceding experiments, and the ants began to clean them. At 8.30 they were all right. 8.45 ditto. At 9 they had been completely cleaned, so that I could not distinguish them; but there was no fighting going on in the nest.

On the same day, at 9.45 A.M., I put into nest 64 two more as before. At 10 they were both quite at home among the other ants. 10.15 ditto, 10.30 ditto, 11 ditto, 12 ditto, 1 ditto. I then put in a stranger; and she was at once fiercely attacked.

September 8.—Put in two more of the ants which had emerged from the pupæ, as before, at 9.30 A.M. At 9.45 they were all right. 10 ditto, 10.30 ditto, 11 ditto, 11.30 ditto, 12 ditto, 1 ditto.

On the other hand, on September 14, I put one of these ants in the same manner into nest No. 60 at 6.30 A.M. She was at once attacked. At 6.45 she was being dragged about by an antenna. 7 ditto. At 7.30 she was by herself in one corner. At 8.30 she was again being dragged about. 9.30 ditto. The difference, therefore, was unmistakable.

Once more, on July 29 I put some pupæ of *Formica fusca* from out of doors under the charge of three ants from nest No. 36.

August 3.—Several had come out, and I put two of them into the nest of their nurses (No. 36) at 2 P.M. Both were at once attacked. At 2.45 they were being dragged about. 3 ditto. 3.30 one was being dragged about. 4, both were being attacked. Eventually one was turned out of the nest. The other I lost sight of.

August 4.—Put two more of this batch into nest No. 36, at 12.30. One was at once attacked. 1, one was being dragged about by an antenna. 2.30, both were being attacked. At 2.45 one was dragged out of the nest.

I then put back one of the old ones; as might have been expected, she was received quite amicably.

I then tried the same experiment with another species, *Lasius niger*. I took some pupæ from two of my nests, which I knew not to be on friendly terms, and which I will call 1 and 2, and confided each batch to three or four ants taken from the other nest. When they had come to maturity I introduced them into the nests as before.

They were taken from their nest on September 20; and the results were as follows.

Pupæ from nest 1 confided to ants from nest 2.

September 20.—Put one of the young ones into nest 2 at 7.15 A.M. Several at once threatened her. At 7.25 one of the ants seized her by an antenna, and began dragging her about. 7.30, she was still being dragged about. 8, ditto. 8.15, she was now being dragged about by three ants. 8.30, she was still attacked. 9, ditto. At 9.15 she was dragged out of the nest.

September 23.—Put two of the young ants into nest 1 at 9.15 A.M. One was at once attacked, and the other a few minutes afterwards. 9.45, both were attacked. 10, ditto. One was now dead and hanging on to a leg of assailant. 10 15 ditto. 10.45, both were still being dragged about.

At 11 A.M. I put into nest 2 three more very young

ones. At 11.10 one was attacked. At 11.20 all three were being viciously attacked, and yet one was nearly cleaned. At 12 one was being attacked, one was alone in a corner, the other we could not find. At 12.10 one was dragged out of the nest and then abandoned, on which, to my surprise, she ran into the nest again, which no old ant would have done. She was at once again seized by an antenna. At 12.30 she was still being dragged about; the second was being cleaned. In this instance, therefore, I think two out of the three were eventually accepted as inmates of the nest.

September 25.—Put two of the young ones into nest 1 at 2.30 P.M. At 2.45 one was attacked, but not viciously. 3 ditto, 3.15 ditto. No notice was taken of the other, though several ants came up and examined her. 3.30, the first was not attacked, the second was almost cleaned. 4, the first has been again attacked, but not viciously, and moreover has been partly cleaned. The second was evidently received as a friend, and was almost cleaned. 4.30, they are both comfortably among the others and are almost clean. At 5 I could no longer distinguish them.

I now pass to the other batch, namely, pupæ from nest 2 with ants from nest 1.

September 25.—Put three of the young ants into nest 1 at 9.30 A.M. At 9.45 two were attacked, the third was by herself. 10 ditto. At 10.15 one made her escape from the nest. At 10.20 the third was attacked. At 10.30 one of them was dragged out of the nest, and then abandoned. At 10.50 the third also was dragged out of the nest.

I then put two of these ants and a third young one into nest 2. At 11.15 A.M. they seemed quite happy; but at 11.30 two were being dragged about; the third, who was very young, was, on the contrary, being carefully cleaned. At 12 this last one was undistinguishable; of the other two, one was being attacked, the second was taken no notice of, though several ants came up to her. At 12.5 the first was dragged out of the nest and then abandoned; the second was being carefully cleaned. This went on till 12.20, when the paint was entirely removed.

September 27.—I put in three more of these young ants into nest 1, at 7.45 A.M. At 8 o'clock they seemed quite at home among the other ants. A few minutes after, one was being held by a leg; the other two seemed quite at home. At 8.30 one was almost cleaned, the other I could not see. At 9 two of them were quite at home, but I could not see the third. At 9.30 they were both nearly cleaned; and after that we were no longer able to distinguish them.

Thinking the results might be different if the ants were allowed to become older before being returned into their nests, I made no further observations with these ants for two months. I then took two of the ants which had emerged from the pupæ separated on September 20, and which had been brought up by ants from nest 2, and on November 22 I put them back at 12 A.M. in their old nest (that is to say, in nest 1), having marked them as usual, with paint. They showed no signs of fear, but ran about among the other ants with every appearance of being quite at home. At 12.15 ditto. At 12.30 one was being cleaned. At 12.45 both were being cleaned; and by 1 o'clock they could scarcely be distinguished from the other ants. There had not been the slightest symptoms of hostility. After this hour we could no longer identify them; but the nest was carefully watched throughout the afternoon, and I think I can undertake to say that they were not attacked. When we left off watching, the nest was enclosed in a box. The next morning I examined it carefully, to see if there were any dead bodies. This was not the case; and I am satisfied, therefore, that neither of these two ants was killed. To test these ants, I then, on November 24, at 8.30 A.M., put into the nest two ants from nest 2. At 8.40 one was attacked; the other had hid herself away in a corner. At 9.15 both of the ants were being dragged about. At 9.35 one was dragged out of the nest and then released, and the other a few minutes afterwards. After watching them for some time to see that they remained outside, I restored them to their own nest. The contrast, therefore, was very marked.

Again, on November 25, I took two ants which had emerged from pupæ belonging to nest 2, removed on September 20, and brought up by ants from nest 1, and put them back into their old nest at 2 P.M. They were watched continuously until 4 P.M., but were not attacked, nor even threatened. The following morning one of them was quite well, the other one we could not distinguish; she had probably been cleaned. If she had been killed, we must have found her dead body. I then at 10 A.M. put in two more. At 10.30 one of them was attacked for a moment, but only for a moment. With this exception neither of them was attacked until 2 o'clock, when one of them was again seized and dragged about for a minute or two, but then released again. We continued watching them till half-past 4, when they seemed quite at home amongst the others. On the other hand a stranger, put in as a test at 12, was at once attacked. It was curious, however, that although she was undoubtedly attacked, yet at the very same time another ant began to clean her.

The next morning we found one ant, and only one, in the box outside the nest; and this turned out to be the stranger of yesterday. She had been almost cleaned; but there were one or two small particles of paint still remaining, so that there could be no doubt of her identity.

The next day, November 27, I put in three more of the ants derived from these pupæ at 10 A.M. At 10.30 they were all right, running about amongst the others. At 11 o'clock the same was the case; but whilst I was looking again shortly afterwards, one of them was seized by an antenna and dragged a little way, but released again in less than a minute. Shortly afterwards one of the others was also seized, but let go again almost immediately. At one o'clock they were all right, and also at two. They had, however, in the meantime been more than once threatened, and even momentarily seized, though they were never dragged about as strangers would have been. At three o'clock I found one of them dead; but I think I must have accidentally injured her, and I do not believe that she was killed by the other ants, though I cannot speak quite positively about it. The other two were quite at home, and had been partly cleaned. At six one of them was running about comfortably amongst the rest; the other I could not distinguish; but certainly no ant was being attacked.

November 28.—I put in the last two ants from the above-mentioned batch of pupæ at noon. Like the preceding, these ants were occasionally threatened, and even sometimes attacked for a moment or two; but the other ants soon seemed to find out their mistake, and on the whole they were certainly treated as friends, the attacks never lasting more than a few moments. One of them was watched at intervals of half an hour until 5 P.M.; the other we could not distinguish after 3 P.M., the paint having been licked off; but we should certainly have observed it had she been attacked.

On the whole, then, all the thirty-two ants belonging to *Formica fusca* and *Lasius niger*, removed from their nest as pupæ, attended by friends and restored to their own nest, were amicably received.

What is still more remarkable, of twenty-two ants belonging to *F. fusca*, removed as pupæ, attended by strangers, and returned to their own nest, twenty were amicably received. As regards one I am doubtful; the last was crippled in coming out of the pupa-case; and to this perhaps her unfriendly reception may have been due.

Of the same number of *Lasius niger* developed in the same manner from pupæ tended by strangers belonging to the same species, and then returned into their own nest, nineteen were amicably received, three were attacked, and about two I feel doubtful.

On the other hand, fifteen specimens belonging to the same two species, removed as pupæ, tended by strangers belonging to the same species, and then put into the strangers' nest, were all attacked.

The results may be tabulated as follows:—

Pupæ brought up by friends and replaced in their own nest.	Pupæ brought up by strangers.	
	Put in own nest.	Put in strangers' nest.
Attacked 0	7[1]	15
Received amicably . . 33	37	0

The differences cannot be referred to any difference of temperament in different nests. The specimens of *F. fusca* experimented with in August and September last were taken principally from two nests, numbered respectively 36 and 60. Now, while nest 36, in most

---

[1] I do not feel sure about three of these.

cases, amicably received ants bred from its own pupæ but tended by ants from 60, it showed itself fiercely hostile to ants from pupæ born in nest 60, even when these had been tended by ants from nest 36. Nest 60, again, behaved in a similar manner; amicably receiving, as a general rule, its own young, even when tended by ants from 36; and refusing to receive ants born in nest 36, even when tended by specimens from nest 60.

These experiments seem to indicate that ants of the same nest do not recognise one another by any password. On the other hand, they seem to show that if ants are removed from a nest in the pupa-state, tended by strangers, and then restored, some at least of their relatives are puzzled, and for a time doubt their claim to consanguinity. I say some, because while strangers under the circumstances would have been immediately attacked, these ants were in every case amicably received by the majority of the colony, and it was sometimes several hours before they came across one who did not recognise them.

In all these experiments, however, the ants were taken from the nest as pupæ, and though I did not think the fact that they had passed their larval existence in the nest could affect the problem, still it might do so. I determined therefore to separate a nest before the young were born, or even the eggs laid, and then ascertain the result. Accordingly I took one of my nests of *F. fusca*, which I began watching on Sept. 13, 1878, and which contained two queens, and on February 8,

1879, divided it into halves, which I will call A and B, so that there were approximately the same number of ants with a queen in each division. At this season, of course, the nest contained neither young nor even eggs. During April both queens began to lay eggs. On July 20 I took a number of pupæ from each division and placed each lot in a separate glass, with two ants from the same division. On August 30 I took four ants from the pupæ bred in B, and one from those in A (which were not quite so forward), and after marking them as usual with paint, put the B ants into nest A, and the A ant into nest B. They were received amicably and soon cleaned. Two, indeed, were once attacked for a few moments, but soon released. On the other hand, I put two strangers into nest A, but they were at once driven out. For facility of observation I placed each nest in a closed box. On the 31st I carefully examined the nests and also the boxes in which I placed them. I could only distinguish one of the marked ants, but there were no dead ants either in the nests or boxes.

I carefully examined the box in the same way for several successive mornings, but there was no dead ant. If there had been I must have found the body, and I am sure, therefore, that these ants were not attacked.

Again, on August 31 I put two more of the ants which had emerged from the pupæ taken out of nest B, and nursed by ants from that nest, into nest A at

10 A.M. At 10.30 A.M. they were quite comfortable amongst the others. At 11 A.M. I looked again and they seemed quite at home, as also at 11.30 A.M., after which I looked every hour, and they were never attacked. The next morning I found them peaceably among the other ants.

On September 15 I put three of the ants which had emerged from the pupæ taken out of nest A, and nursed by ants from that nest, and put them into nest B at 1.30 P.M. They seemed to make themselves quite at home. I looked again at 2.30 P.M., with the same result. At 3.30 P.M. I could only find two, the third having no doubt been cleaned, but no ant was being attacked. At 5.30 P.M. they were no longer distinguishable, but if any one was being attacked we must have seen it. The next morning they all seemed quite peaceful, and there was no dead ant in the box. I looked again on the 17th and 19th, but could not distinguish them. As, however, there was no dead ant, they certainly had not been killed. I then put in a stranger; she was soon attacked and driven out of the nest—showing that, as usual, they would not tolerate an ant whom they did not recognise as in some way belonging to the community.

Again, on April 10, 1881, I divided a two-queened nest of *Formica fusca*, leaving a queen in each half. At that time no eggs had yet been laid, and of course there were no larvæ or pupæ. In due course both queens laid eggs, and young ants were brought up in each half of the nest. I will call the two halves as before A and B.

On August 15, at 9 A.M., I put three of the young ants from A into B, and three from B into A. At 9.30 A.M. none were attacked, 10 A.M. ditto, 10.30 A.M. ditto. One was being cleaned; 12 A.M. ditto, 2 P.M. ditto. In fact, they seemed quite at home with the other ants. The next morning I was unable to recognise them, the paint having been entirely removed. The ants were all peaceably together in the nest, and there were no dead ones either in the nest or in the outer box. It is evident, therefore, that they had been treated as friends.

August 17.—I put in three more from B into A at noon. At 12.30 P.M. they were with the other ants; at 1 P.M., ditto, at 2 P.M. ditto, at 3 P.M. ditto, at 5 P.M. ditto. The following morning I was still able to recognise them, though most of the paint had been removed. They also were evidently treated as part of the community.

September 19.—Put in three more from A into B at 8.30 A.M. I looked at them at intervals of half an hour, but none of them were attacked. Next morning there was no ant outside the nest, nor had any been killed.

October 10.—Put in three more at 7 A.M., and looked at intervals of an hour. They were not attacked, and evidently felt themselves among friends. The next morning I was still able to recognise two,

There was no dead ant either in the nest or the outer box.

Lastly, on October 15, I put in four more at 7 A.M., and watched them all day at short intervals. They exhibited no sign of fear, and were never attacked. In fact, they made themselves quite at home, and were evidently, like the preceding, recognised as friends. For the sake of comparison at noon I again put in a stranger. Her behaviour was in marked contrast. The preceding ants seemed quite at home, walked about peaceably among the other ants, and made no attempt to leave the nest. The stranger, on the contrary, ran uneasily about, started away from any ant she met, and made every effort to get out of the nest. After she had three times escaped from the nest, I put her back with her own friends.

Thus, then, when a nest of *Formica fusca* was divided early in spring, and when there were no young, the ants produced in each half were in twenty-eight cases all received as friends. In no case was there the slightest trace of enmity.

These observations seem to me conclusive as far as they go, and they are very surprising. In the previous experiments, though the results were similar, still the ants experimented with had been brought up in the nest, and were only removed after they had become pupæ. It might therefore be argued that the ants having nursed them as larvæ, recognized them when they came to maturity; and though this would certainly be in the highest degree improbable, it could not be said to be impossible. In the present case, however, the old ants had absolutely never seen the young ones until the moment when, some days after arriving at maturity, they were introduced into the nest; and yet in twenty-one cases they were undoubtedly recognised as belonging to the community.

It seems to me, therefore, to be established by these experiments that the recognition of ants is not personal or individual; that their harmony is not due to the fact that each ant is individually acquainted with every other member of the community.

At the same time, the fact that they recognise their friends even when intoxicated, and that they know the young born in their own nest even when they have been brought out of the chrysalis by strangers, seems to indicate that the recognition is not effected by means of any sign or password

CHAPTER VII.

POWER OF COMMUNICATION.

THE Social Hymenoptera, according to Messrs. Kirby and Spence,[1] 'have the means of communicating to each other information of various occurrences, and use a kind of language which is mutually understood, ... and is not confined merely to giving intelligence of the approach or absence of danger; it is also co-extensive with all their other occasions for communicating their ideas to each other.'

Huber assures us as regards Ants[2] that he has 'frequently seen the antennæ used on the field of battle to intimate approaching danger, and to ascertain their own party when mingled with the enemy; they are also employed in the interior of the ant-hill to apprise their companions of the presence of the sun, so favourable to the development of the larvæ, in their excursions and emigrating to indicate their route, in their recruitings to determine the time of departure,' &c. Elsewhere also he says[3] 'that should an Ant fall in with any of her associates from the nest they put her in the right way by the contact of their antennæ.'

[1] *Introduction to Entomology*, ii. p. 50.   [2] *Loc. cit.* p. 206.
[3] *Loc. cit.* p 157.

These statements are most interesting; and it is much to be regretted that he has not given us in detail the evidence on which they rest. In another passage, indeed, he himself says,[1] 'If they have a language, I cannot give too many proofs of it.' Unfortunately, however, the chapter which he devotes to this important subject is very short, and occupied with general statements rather than with the accounts of the particular experiments and observations on which those statements rest. Nor is there any serious attempt to ascertain the nature, character, and capabilities of this antennal language. Even if by motions of these organs Ants and Bees can caress, can express love, fear, anger, &c., it does not follow that they can narrate facts or describe localities.

The facts recorded by Kirby and Spence are not more explicit. It is therefore disappointing to read in the chapter especially devoted to this subject, that, as regards the power possessed by Ants and Bees to communicate and receive information, 'it is only necessary to refer you to the endless facts in proof, furnished by almost every page of my letters on the history of Ants and of the Hive Bee. I shall therefore but detain you for a moment with an additional anecdote or two, especially with one respecting the former tribe, which is valuable from the celebrity of the narrator.'

The first of these anecdotes refers to a Beetle (*Ateuchus pilularius*) which, having made for the

---

[1] *Loc. cit.* p. 205.

reception of its eggs a pellet of dung too heavy for it to move, 'repaired to an adjoining heap and soon returned with three of his companions. All four now applied their united strength to the pellet, and at length succeeded in pushing it out, which being done, the three assistant Beetles left the spot and returned to their own quarters.' This observation rests on the authority of an anonymous German artist; and though we are assured that he was a 'man of strict veracity,' I am not aware that any similar fact has been recorded by any other observer. I am by no means satisfied that his explanation of what took place is correct. M. Fabre,[1] in his interesting observations, places the facts in a very different light.

The second case is related by Kalm, on the authority of Dr. Franklin, but again does not seem to me to justify the conclusions drawn from it by Messrs. Kirby and Spence. Dr. Franklin having found a number of ants in a jar of treacle, shook them out and suspended the jar ' by a string from the ceiling. By chance one ant remained, which, after eating its fill, with some difficulty found its way up the string, and, thence reaching the ceiling, escaped by the wall to its nest. In less than half an hour a great company of ants sallied out of their hole, climbing the ceiling, crept along the string into the pot and began to eat again; this they continued until the treacle was all consumed, one swarm running up the string while another passed

[1] *Souvenirs Entomologiques.*

down. It seems indisputable that the one ant had in this instance conveyed news of the booty to his comrades, who would not otherwise have at once directed their steps in a body to the only accessible route.'[1]

Elsewhere, Messrs. Kirby and Spence say:[2]—'If you scatter the ruins of an ants' nest in your apartment, you will be furnished with another proof of their language. The ants will take a thousand different paths, each going by itself, to increase the chance of discovery; they will meet and cross each other in all directions, and perhaps will wander long before they can find a spot convenient for their reunion. No sooner does any one discover a little chink in the floor through which it can pass below than it returns to its companions, and, by means of certain motions of its antennæ, makes some of them comprehend what route they are to pursue to find it, sometimes even accompanying them to the spot; these, in their turn, become the guides of others, till all know which way to direct their steps.'

Here, however, Messrs. Kirby and Spence do not sufficiently distinguish between the cases in which the ants were guided, from those in which they were directed to the place of safety. It is obvious, however, that the power of communication implied in the latter case is much greater than in the former.

A short but very interesting paper by Dujardin on this subject is contained in the 'Annales des Sciences' for 1852. He satisfied himself that some bees which

---

[1] *Loc. cit.* p. 422.   [2] *Introd. to Entomology*, vol. ii **p. 6.**

came to honey put out by him for the purpose 'avaient dû recevoir dans la ruche un avertissement porté par quelques-unes de celles qui étaient venues isolément, soit à dessein, soit par hasard.' That no doubt might remain, he tried the following experiment, which he says, 'me parait tout-à-fait concluante. Dans l'épaisseur d'un mur latéral à 18 mètres de distance des ruches A et B, se trouve une niche pratiquée, suivant l'usage du pays, pour constater la mitoyenneté, et recouverte par un treillage et par une treille, et cachée par diverses plantes grimpantes. J'y introduisis, le 16 novembre, une soucoupe avec du sucre légèrement humecté ; puis j'allai présenter une petite baguette enduite de sirop à une abeille sortant de la ruche. Cette abeille s'étant cramponnée à la baguette pour sucer le sirop, je la transportai dans la niche sur le sucre, où elle resta cinq ou six minutes jusqu'à ce qu'elle se fut bien gorgée ; elle commença alors à voler dans la niche, puis deçà et delà devant le treillage, la tête toujours tournée vers la niche, et enfin elle prit son vol vers la ruche, où elle rentra.

'Un quart d'heure se passa sans qu'il revînt une seule abeille à la niche ; mais, à partir de cet instant, elles vinrent successivement au nombre de trente, explorant la localité, cherchant l'entrée de la niche qui avait dû leur être indiquée, et où l'odorat ne pouvait nullement les guider, et, à leur tour vérifiant avant de retourner à la ruche, les signes qui leur feraient retrouver cette précieuse localité ou qui leur permettraient de l'indiquer à d'autres. Tous les jours suivants les abeilles de la ruche A vinrent plus nombreuses à la niche où j'avais soin de renouveler le sucre humecté, et pas une seule de la ruche B n'eut le moindre soupçon de l'existence de ce trésor et ne vint voler de ce côté. Il était facile, en effet, de constater que les premières se dirigeaient exclusivement de la ruche à la niche, et réciproquement.'

It is of course clear from these observations that the ants and bees accompanied their fortunate friends to the stores of food which they had discovered, but this really does not in itself imply the possession of any great intelligence.

That ants and bees have a certain power of communication cannot, indeed, be doubted. Several striking cases are mentioned by M. Forel. For instance, on one occasion an army of Amazon ants (*Polyergus rufescens*) was making an expedition to attack a nest of *F. rufibarbis*. They were not, however, quite acquainted with the locality. At length it was discovered:—'Aussitôt,' he observes, ' un nouveau signal fût donné, et toutes les amazones s'élancèrent dans cette direction.' On another occasion he says:— ' Je mis un gros tas de *T. cæspitum* d'une variété de grande taille à un décimètre d'un des nids d'une colonie de Pheidole pallidula. En un clin d'œil l'alarme fût répandue, et des centaines de Pheidole se jetèrent au-devant de l'ennemi.'

The species of *Camponotus*, when alarmed, 'non

seulement se frappent vivement et à coups répétés les uns les autres, mais en même temps ils frappent le sol deux ou trois fois de suite avec leur abdomen, et répètent cet acte à de courts intervalles, ce qui produit un bruit très marqué qu'on entend surtout bien lorsque le nid est dans un tronc d'arbre.'[1]

It would even seem, according to M. Forel, that some species understand the signs of others. Thus *F. sanguinea*, he says,[2] is able to seize ' l'instant où les pratensis se communiquent le signal de la déroute, et elles savent s'apprendre cette découverte les unes aux autres avec une rapidité incroyable. Au moment même où l'on voit les pratensis se jeter les unes contre les autres en se frappant de quelques coups rapides, puis cesser toute résistance et s'enfuir en masse, on voit aussi les *sanguinea* se jeter tout-à-coup au milieu d'elles sans la plus petite retenue, mordant à droite et à gauche comme des *Polyergus*, et arrachant des cocons de toutes les *pratensis* qui en portent.'

M. Forel is of opinion (p. 364) that the different species differ much in their power of communicating with one another. Thus, though *Polyergus rufescens* is rather smaller than *F. sanguinea*, it is generally victorious, because the ants of this species understand one another more quickly than those of *F. sanguinea*.

These statements are extremely interesting, and certainly appear to imply considerable intelligence. If, however, his inferences are correct, and the social

[1] *Loc. cit.* p. 355. [2] *Loc. cit.* p. 359.

Hymenoptera are really so highly gifted, it ought not to be necessary for us to rely on accidental observations; we ought to be able to test them by appropriate experiments.

Those which I have made with reference to bees will be described in a subsequent chapter.

Every one knows that if an ant or a bee in the course of her rambles has found a supply of food, a number of others will soon make their way to the store. This, however, does not necessarily imply any power of describing localities. A very simple sign would suffice, and very little intelligence is implied, if the other ants merely accompany their friend to the treasure which she has discovered. On the other hand, if the ant or bee can describe the locality, and send her friends to the food, the case is very different. This point, therefore, seemed to me very important; and I have made a number of observations bearing on it.

The following may be taken as a type of what happens under such circumstances. On June 12, 1874, I put a *Lasius niger*, belonging to a nest which I had kept two or three days without food, to some honey. She fed as usual, and then was returning to the nest, when she met some friends, whom she proceeded to feed. When she had thus distributed her stores, she returned alone to the honey, none of the rest coming with her. When she had a second time laid in a stock of food, she again in the same way fed several ants on her way towards the nest; but this time five of those so fed returned with her to the honey. In due course these five would no doubt have brought others, and so the number at the honey would have increased.

Some species, however, act much more in association than others—*Lasius niger*, for instance, much more than *Formica fusca*.

In March 1877 I was staying at Arcachon. It was a beautiful and very warm spring day, and numerous specimens of *Formica fusca* (Pl. I, fig. 3) were coursing about on the flagstones in front of our hotel. At about 10.45 A.M. I put a raisin down before one of them. She immediately began licking it, and continued till 11.2 A.M., when she went off almost straight to her nest, the entrance to which was about twelve feet away. In a few minutes she came out again, and reached the fruit, after a few wanderings, at about 11.18 A.M. She fed till 11.30 A.M., when she returned once more to the nest.

At 11.45 another ant accidentally found the fruit. I imprisoned her.

At 11.50 the first returned, and fed till 11.56, when she went off to the nest. On the way she met and talked with three ants, none of whom, however, came to the fruit. At 12.7 she returned, again alone, to the fruit.

On the following day I repeated the same experiment. The first ant went backwards and forwards between the raisin and the nest for several hours, but only six others found their way to it.

The details of this observation will be found in the Appendix.

Again, on July 11, 1875, I put out some pupæ in a saucer, and at 5.55 p.m. they were found by a *F. fusca*, who as usual carried one off to the nest.

At 6 p.m. she returned and took another. Again

6. 1	,,	,,
6. 3	,,	,,
6. 4	,,	,,
6. 5	,,	,,
6. 6	,,	,,
6. 7	,,	,,
6. 8	,,	,,
6. 9	,,	,,
6.10	,,	,,
6.11	,,	,,
6.12	,,	,,
6.14	,,	,,
6.15	,,	,,
6.16	,,	,,
6.17	,,	,,
6.19	,,	,,
6.20	,,	,,
6.21	,,	,,
6.23	,,	,,
6.25	,,	,,
6.27	,,	,,
6.29	,,	,,
6.30	,,	,,

At 6.31 P.M. she returned and took another. Again

6.33	,,	,,
6.35	,,	,,
6.36	,,	,,
6.37	,,	,,
6.38	,,	,,
6.40	,,	,,
6.41	,,	,,
6.45	,,	,,
6.47	,,	,,
6.49	,,	,,
6.50	,,	,,
6.51	,,	,,
6.52	,,	,,
6.53	,,	,,
6.55	,,	,,
6.56	,,	,,
6.57	,,	,,
7. 0	,,	,,
7. 1	,,	,,
7. 2	,,	,,
7. 6	,,	,,

After these 45 visits, she came no more till 8 P.M.; but when I returned at 10 P.M. I found all the pupæ gone. During the time she was watched, however, she brought no other ant to assist.

I also made similar experiments with *Myrmica ruginodis* and *Lasius niger*, imprisoning (as before) all ants that came, except the marked ones, and with similar results. The details will be found in the Appendix, but need not be given in full here.

I then tried the following experiment:—

In figure 3, A is the ants' nest, o the door of the nest. M is the section of a pole on which the whole apparatus is supported. B is a board 2 feet long; C, D, E, and F are slips of glass connected with the board B by narrow strips of paper G, H, I. K is a movable strip of paper, 1½ inch long, connecting the glass F with the strip H; and L is another movable strip of paper, as nearly as possible similar, connecting H and I. On each of the slips of glass C and F I put several hundred larvæ of *L. flavus*. The object of the larvæ on C was to ascertain whether, under such circumstances, other ants would find the larvæ accidentally; and I may say at once that none did so. I then put an ant (A), whom I had imprisoned overnight, to the larvæ on F. She took one, and, knowing her way, went straight home over the bridge K and down the strip H. Now it is obvious that by always causing the marked ant (A) to cross the bridge K on a particular piece of paper, and if at other times the papers K and L were reversed, I should be able to ascertain whether other ants who came to the larvæ had had the direction and position explained to them; or whether, having only

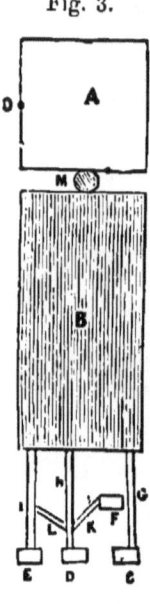

Fig. 3.

POWERS OF COMMUNICATION. 165

been informed by A of the existence of the larvæ, they found their way to them by tracking A's footsteps. If the former, they would in any case pass over the bridge K by whichever strip of paper it was constituted. On the other hand, if they found the larvæ by tracking, then as the piece of paper by which A passed was transferred to L, it would mislead them and carry them away from the larvæ to I. In every case, then, I transposed the two papers forming the little bridges as soon as the ant A had crossed over K and L.

I put her (November 7, 1875) to the larvæ on F at 6.15 A.M. After examining them carefully, she returned to the nest at 6.34. No other ants were out; but she at once reappeared with four friends and reached the larvæ at 6.38. None of her friends, however, crossed the bridge; they went on to D, wandered about, and returned home. A returned to the larvæ at 6.47, this time with one friend, who also went on to D and returned without finding the larvæ.

7. 0.	Ant A to larvæ.			
7. 8	,,		An ant at 7.10	{went over L to I.
7.17	,,	with a friend, who at	7.21	,,
7.25	,,	{with two friends, one of whom at}	7.27	,,
7.32	,,	the other at	7.35	,,
7.39	,,	{with a friend who went on to D, and then at ......}	7.41	,,

7.46	Ant A to larvæ.	An ant at 7.42	{went over L to I.	
7.55	,,	,,	7.47	,,
8. 3	,,	,,	7.48	,,
8. 8	,,	,,	7.54	,,
8.19	,,	,,	7.57	,,
8.24	,,	,,	9.10	found the larvæ.
8.39	,,	,,	9.30	went over L to I.
8.50	,,			
9.12	,,			
9.22	,,			
9.40	,,			
9.47	,,			
9.55	,,			
10.35	,,			

At 10.35 I imprisoned her till 12.30, when I put her again to the larvæ.

12.48	back to larvæ.					
12.55	,,	An ant at 12.58 went over L to I.				
1. 0	,,	,,	1. 1	,,	,,	
1.15	,,	,,	1.10	,,	,,	
1.20	,,	,,	1.13	,,	,,	

After this she did not come any more. During the time she made, therefore, 25 visits to the larvæ; 21 other ants came a distance of nearly 4 feet from the nest and up to the point of junction within 2 inches of the larvæ; but only one passed over the little bridge to the larvæ, while 15 went over the bridge L to I. On repeating this experiment with another marked ant, she herself made 40 journeys, during which 19 other ants found their way to the point of junction. Only 2 went over the little bridge to the larvæ, 8 went over L to I, and the remainder on to D.

In another similar experiment the marked ant made 16 journeys; and during the same time 13 other ants came to the point of junction. Of these 13, 6 went on to D, 7 crossed over L to I, and not one found the larvæ. Thus altogether, out of 92 ants, 30 went on to D, 51 crossed over in the wrong direction to I, and only 11 found their way to the larvæ.

From January 2 to January 24 (1875) I made a series of similar observations; and during this time 56 ants came in all. Of these, 20 went straight on to D, 26 across the paper to I, and only 10 to the larvæ.

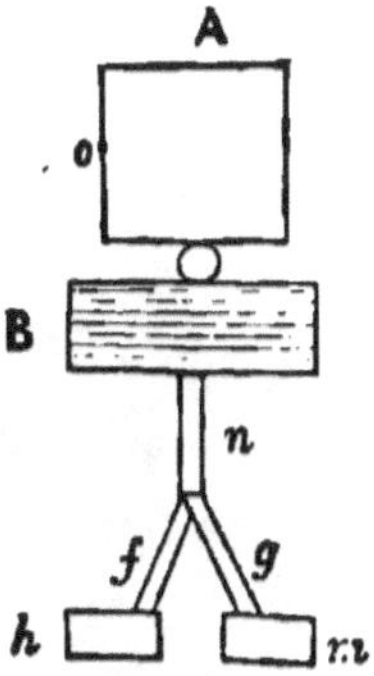

This, I think, gives strong reason to conclude that, under such circumstances, ants track one another by scent.

Fig. 4.

I then slightly altered the arrangement of the papers as shown in the accompanying diagram (fig. 4). A, as before, is the nest, o being the door. B is the board; h is a glass on which are placed the larvæ; m is a similar glass, but empty; n a strip of paper: to the end of n are pinned two other strips *f* and *g*, in such a manner that they can be freely turned round, so that each can be turned at will either to *h* or *m*. Under ordinary circumstances the paper *f*, as in the figure, was turned to the larvæ; but whenever any ant, excepting the marked one, came, I turned the papers, so that *f* led to *m* and *g* to *h*. The result was striking, and I give the observation in full in the Appendix. In all, 17 ants came, every one of whom took the wrong turn and went to *m*.

Although the observations above recorded seem to me almost conclusive, still I varied the experiments once more (see fig. 5), making the connexion between the board B and the glass containing the larvæ by three separate but similar strips of paper, *d*, *e*, and *f*, as shown in the figure. Whenever, however, a strange ant came, I took up the strip *f* and rubbed my finger over it two or three times so as to remove any scent, and then replaced it. As soon as the stranger had reached the paper *e*, I took up the strip *d*, and placed it so as to connect *e* with the empty glass *m*. Thus I escaped the necessity of changing the paper *f*, and yet had a scented bridge between *e* and *m*. The details, as before, are given in the Appendix.

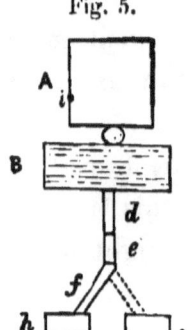

Fig. 5.

In this experiment the bridge over which the marked ant passed to the larvæ was left in its place, the scent, however, being removed or obscured by the friction of my finger; on the other hand, the bridge (*d*) had retained the scent, but was so placed as to lead away from the larvæ; and it will be seen that, under these circumstances, out of 41 ants which found their way towards the larvæ as far as *e*, 14 only passed over the bridge *f* to the larvæ, while 27 went over the bridge *d* to the empty glass *m*.

Taking these observations as a whole, 150 ants came to the point *e*, of which 21 only went on to the larvæ, while 95 went away to the empty glass. These experiments, therefore, seem to show that when an ant has discovered a store of food and others flock to it, they are guided in some cases by sight, while in others they track one another by scent.

I then varied the experiment as follows:—I put an ant (*L. niger*) to some larvæ as usual, and when she knew her way, I allowed her to go home on her own legs; but as soon as she emerged from the nest, if she had any friends with her, I took her up on a bit of paper and carried her to the larvæ. Under these circumstances very few ants indeed found their way to them. Thus, on June 23, 1876, at 5.30, an ant which had been previously under observation was put to some larvæ. She took one and returned as usual to the nest. At 5.34 she came out with no less than 10 friends, and was then transferred to the larvæ. The others wandered about a little, but by degrees returned to the nest, not one of them finding their way to the larvæ. The first ant picked up a larva, returned, and again came out of the nest at 5.39 with 8 friends, when exactly the same thing happened. She again came out with companions at the undermentioned times:—

Hour.	Number of Friends.	Hour.	Number of Friends
5.44	4	6.44	—
5.47	4	6.46	3
5.49	—	6.49	2
5.52	—	6.56	—
5.54	5	6.59	—
5.57	2	7. 2	2
5.59	2	7. 4	—
6. 1	5	7. 6	3
6. 4	1	7. 8	3
6. 7	—	7.10	5
6.11	3	7.13	—
6.14	4	7.17	3
6.17	6	7.19	7
6.20	—	7.21	5
6.23	5	7.24	—
6.25	6	7.26	3
6.29	8	7.29	1
6.32	2	7.31	2
6.35	—	7.35	—
6.42	4		

Thus during these two hours more than 120 ants came out of the nest in company with the one under observation. She knew her way perfectly; and it is clear that if she had been left alone, all, or at least most of, these ants would have accompanied her to the store of larvæ. Three of them were accidentally allowed to do so; but of the remainder, only 5 found their way to the larvæ; all the others, after wandering about a while, returned hopelessly to the nest.

One of the ants which I employed in my experiments was under observation several days. I was, however, away from home most of the day, and when I left in the morning and went to bed at night I put her in a bottle; but the moment she was let out she began to work again. On one occasion I was away for a week, and on my return I let her out of the bottle, placing her on a little heap of larvæ about 3 feet from the nest. Under these circumstances I certainly did not expect her to return. However, though she had thus been six days in confinement, the brave little creature immediately picked up a larva, carried it off to the nest, and, after half an hour's rest, returned for another.

I conclude, then, that when large numbers of ants come to food they follow one another, being also to a certain extent guided by scent. The fact, therefore, does not imply any considerable power of intercommunication. There are, moreover, some other circumstances which seem to show that their powers in this respect are but limited. For instance, I have already mentioned that if a colony of *Polyergus* changes the situation of its nest, the mistresses are all carried to the new one by the slaves. Again, if a number of *F.*

fusca are put in a box, and in one corner a dark place of retreat is provided for them with some earth, one soon finds her way to it. She then comes out again, and going up to one of the others, takes her by the jaws. The second ant then rolls herself into a heap, and is carried off to the place of shelter. They then both repeat the same manœuvre with other ants, and so on until all their companions are collected together. Now it seems to me difficult to imagine that so slow a course would be adopted if they possessed any considerable power of descriptive communication.

On the other hand, there can, I think, be no doubt that they do possess some power of the kind.

This seems to me clearly shown by the following observations. In order, if possible, to determine whether the ants in question were brought to the larvæ, or whether they came casually, I tried (1875) the following experiments: I took three tapes, each about 2 feet 6 inches long, and arranged them parallel to one another and about 6 inches apart. One end of each I attached to one of my nests (*L. niger*), and at the other end I placed a small glass. In the glass at the end of one tape I placed a considerable number (300 to 600) of larvæ. In the second I put two or three larvæ only; in the third none at all. The object of the last was to see whether many ants would come to the glasses under such circumstances by mere accident; and I may at once say that but few did so. I then took two ants and placed one of them to the glass with many larvæ, the other to that with two or three. Each of them took a larva and carried it to the nest, returning for another, and so on. After each journey I put another larva in the glass with only two or three larvæ, to replace that which had been removed. Now, if other ants came under the above circumstances as a mere matter of accident, or accompanying one another by chance, or if they simply saw the larvæ which were brought and consequently concluded that they might themselves also find larvæ in the same place, then the numbers going to the two glasses ought to be approximately equal. In each case the number of journeys made by the ants would be nearly the same; consequently, if it was a matter of scent, the two glasses would be in the same position. It would be impossible for an ant, seeing another in the act of bringing a larva, to judge for itself whether there were few or many larvæ left behind. On the other hand, if the friends were brought, then it would be curious to see whether more were brought to the glass with many larvæ, than to that which only contained two or three. I should also mention that, excepting, of course, the marked specimens, every ant which came to the larvæ was imprisoned until the end of the experiment. I give the details in the Appendix.

The results of the above experiments are shown at a glance in the following Table:—

EVIDENCE OF COMMUNICATION.

Tabular View of Experiments on Power of Communication.

Observations	Glass with many larvæ			Glass with one or two larvæ		
	Time occupied	No. of journeys	No. of friends	Time occupied	No. of journeys	No. of friends
	hours			hours		
1	1	7	11			
2	—	—	—	1	6	0
3	—	—	—	2	13	8
4	—	—	—	3	24	5
5	3	38	22	1	10	3
6	2½	32	19			
7	1	5	16			
8	1½	11	21	3	23	2
9	—	—	—	1½	7	3
10	1	15	13	2	21	1
11	2	32	20	1	11	1
12	5	26	10			
13	—	—	—	5	19	1
14	—	—	—	3	20	4
15	2½	41	3	2	5	0
16	1	10	16	2½	10	2
17	4½	53	2	4½	40	10
18	—	—	—	2	20	1
19	1	11	12			
20	—	—	—	1	6	0
21	1½	20	15	4½	74	27
22	—	—	—	1½	25	4
23	4½	71	7			
24	—	—	—	2	35	4
25	2	34	3			
26	1½	35	21	2	18	0
27	2	37	9	1½	15	0
28	1½	9	10	2	14	0
29	2	37	5	1½	25	3
30	1½	9	10	2	14	0
31	2	37	5	1½	25	3
32	2	24	7	1	7	0
33	3½	43	17	3½	26	1
34	1	27	28	1	18	12
35	1	14	2	1	15	9
	52	678	304	59½	545	104

It must be admitted that this mode of observing

is calculated to increase the number of friends brought by the ants to the glass with only 2 or 3 larvæ, for several reasons, but especially because in many cases an ant which had for some time had access to a glass with many larvæ was suddenly deprived of it, and it might well be that some time elapsed before the change was discovered. Some stray ants would, no doubt, in any case have found the larvæ; and we may probably allow for about 25 under this head. Again, some would, no doubt, casually accompany their friends; if we allow 25 also in this respect, we must deduct 50 from each side, and we shall have 254 against 54. Nevertheless, even without any allowances, the results seem to me very definite. Some of the individual cases, especially perhaps experiments 9, 10, 20, 21, and 22 (see Appendix), are very striking; and, taken as a whole, during 52 hours, the ants which had access to a glass containing numerous larvæ brought 304 friends; while during 59 hours those which were visiting a glass with only 2 or 3 larvæ brought only 104 to their assistance.

One case of apparent communication struck me very much. I had had an ant (*L. niger*) under observation one day, during which she was occupied in carrying off larvæ to her nest. At night I imprisoned her in a small bottle; in the morning I let her out at 6.15, when she immediately resumed her occupation. Having to go to London, I imprisoned her again at 9 o'clock. When I returned at 4.40, I put her again to the larvæ. She examined them carefully, but went home without taking one. At this time no other ants were out of the nest. In less than a minute she came out again with 8 friends, and the little troop made straight for the heap of larvæ. When they had gone two-thirds of the way, I again imprisoned the marked ants; the others hesitated a few moments, and then, with curious quickness, returned home. At 5.15 I put her again to the larvæ. She again went home *without a larva*, but, after only a few seconds' stay in the nest, came out with no less than 13 friends. They all went towards the larvæ; but when they got about two-thirds of the way, although the marked ant had on the previous day passed over the ground about 150 times, and though she had just gone straight from the larvæ to the nest, she seemed to have forgotten her way and wandered; and after she had wandered about for half an hour, I put her to the larvæ. Now in this case the 21 ants must have been brought out by my marked one; for they came exactly with her, and there were no other ants out. Moreover, it would seem that they must have been told, because (which is very curious in itself) she did not in either case bring a larva, and consequently it cannot have been the mere sight of a larva which induced them to follow her. I repeated an experiment similar to this more than once.

For instance, one rather cold day, when but few ants were out, I selected a specimen of *Atta testaceo pilosa,* belonging to a nest which I had brought back with me from Algeria. She was out hunting about six feet from home, and I placed before her a large dead bluebottle fly, which she at once began to drag to the nest. I then pinned the fly to a piece of cork, in a small box, so that no ant could see the fly until she had climbed up the side of the box. The ant struggled, of course in vain, to move the fly. She pulled first in one direction and then in another, but, finding her efforts fruitless, she at length started off back to the nest empty-handed. At this time there were no ants coming out of the nest. Probably there were some few others out hunting, but for at least a quarter of an hour no ant had left the nest. My ant entered the nest, but did not remain there; in less than a minute she emerged accompanied by 7 friends. I never saw so many come out of that nest together before. In her excitement the first ant soon distanced her companions, who took the matter with much more *sang-froid,* and had all the appearance of having come out reluctantly, or as if they had been asleep and were only half awake. The first ant ran on ahead, going straight to the fly. The others followed slowly and with many meanderings;.so slowly, indeed, that for twenty minutes the first ant was alone at the fly, trying in every way to move it. Finding this still impossible, she again returned to the nest, not chancing to meet any of her friends by the way. Again she emerged in less than a minute with 8 friends, and hurried on to the fly. They were even less energetic than the first party; and when they found they had lost sight of their guide, they one and all returned to the nest. In the meantime several of the first detachment had found the fly, and one of them succeeded in detaching a leg, with which she returned in triumph to the nest, coming out again directly with 4 or 5 companions. These latter, with one exception, soon gave up the chase and returned to the nest. I do not think so much of this last case, because as the ant carried in a substantial piece of booty in the shape of the fly's leg, it is not surprising that her friends should some of them accompany her on her return; but surely the other two cases indicate a distinct power of communication.

Lest, however, it should be supposed that the result was accidental, I determined to try it again. Accordingly on the following day I put another large dead fly before an ant belonging to the same nest, pinning it to a piece of cork as before. After trying in vain for ten minutes to move the fly, my ant started off home. At that time I could only see two other ants of that species outside the nest. Yet in a few seconds, considerably less than a minute, she emerged with no less than 12 friends. As in the previous case, she ran on ahead, and they followed very slowly and by no means directly, taking, in fact, nearly half an hour to reach the fly. The first ant, after vainly labouring for about a quarter of an hour to move the fly, started off again to the nest. Meeting one of her friends on the way she conversed with her a little, then continued towards the nest, but, after going about a foot, changed her mind, and returned with her friend to the fly. After some minutes, during which two or three other ants came up, one of them detached a leg, which she carried off to the nest, coming out again almost immediately with six friends, one of whom, curiously enough, seemed to lead the way, tracing it, I presume, by scent. I then removed the pin, and they carried off the fly in triumph.

Again, on June 15, 1878, another ant belonging to the same nest had found a dead spider, about the same distance from the nest. I pinned down the spider as before. The ant did all in her power to move it; but after trying for twelve minutes, she went off to the nest. Although for a quarter of an hour no other ant had left the nest, yet in a few seconds she came out again with 10 companions. As in the preceding case, they followed very leisurely. She ran on ahead and worked at the spider for ten minutes; when, as none of her friends had arrived to her assistance, though they were wandering about, evidently in search of something, she started back home again. In three quarters of a minute after entering the nest she reappeared, this time with 15 friends, who came on somewhat more rapidly than the preceding batch, though still but slowly. By degrees, however, they all came up, and after most persevering efforts carried off the spider piecemeal. On July 7, I tried the same experiment with a soldier of *Pheidole megacephala*. She pulled at the fly for no less than fifty minutes, after which she went to the nest and brought five friends exactly as the *Atta* had done.

In the same way, one afternoon at 6.20 I presented a slave of *Polyergus* with a dead fly pinned down. The result was quite different. My ant pulled at the fly for twenty-five minutes, when, as in the previous cases, she returned to the nest. There she remained four or five minutes, and then came out again alone, returned to the fly, and again tried to carry it off. After working fruitlessly for between twenty and twenty-five minutes, she again went back to the nest, staying there four or five minutes, and then returning by herself to the fly once more. I then went away for an hour, but on my return found her still tugging at the fly by herself. One hour later again I looked, with the same result. Shortly afterwards another ant wandering about found the fly, but obviously, as it seemed to me, by accident.

At 3 o'clock on a subsequent day I again put a dead fly pinned on to a bit of cork before a *Formica fusca*, which was out hunting. She tried in vain to carry it off, ran round and round, tugged in every direction, and at length at ten minutes to four she returned to the nest: very soon after she reappeared preceded by one and followed by two friends; these, however, failed to discover the fly, and after wandering about a little returned to the nest. She then set again to work alone, and in about forty minutes succeeded in cutting off the head of the fly, which she at once carried into the nest. In a little while she came out again, this time accompanied by five friends, all of whom found their way to the fly; one of these, having cut off the abdomen of the fly, took it into the nest, leaving three of her companions to bring in the remainder of their prey.

These experiments certainly seem to indicate the possession by ants of something approaching to language. It is impossible to doubt that the friends were brought out by the first ant; and as she returned empty-handed to the nest, the others cannot have been induced to follow her merely by observing her proceedings. In face of such facts as these, it is impossible not to ask ourselves how far are ants mere exquisite automatons; how far are they conscious beings? When we see an ant-hill, tenanted by thousands of industrious inhabitants, excavating chambers, forming tunnels, making roads, guarding their home, gathering food, feeding the young, tending their domestic animals, —each one fulfilling its duties industriously, and without confusion,—it is difficult altogether to deny to them the gift of reason; and the preceding observations tend to confirm the opinion that their mental powers differ from those of men, not so much in **kind** as in degree.

CHAPTER VIII.

ON THE SENSES OF ANTS.

The Sense of Vision.

It is, I think, generally assumed not only that the world really exists as we see it, but that it appears to other animals pretty much as it does to us. A little consideration, however, is sufficient to show that this is very far from being certain, or even probable.

In the case of insects, moreover, the mode of vision is still an enigma. They have, at least many of them have, a large compound eye on each side; and ocelli, generally three in number, situated on the summit of the head. The compound eyes consist of a number of facets, each situated at the summit of a tube, to the base of which runs a fibre of the optic nerve.

The structure of the ocellus and that of the compound eye are essentially different, and it does not seem possible that either the ocellus should be derived from the compound eye, or the compound eye from the ocellus. On the contrary, both seem to point back to a less developed ancestral type. Starting from such an origin, an increase of the separate elements and an improvement of the lens would lead to the ocellus, while

TWO KINDS OF EYES. 183

an increase of the number of eyes would bring us to the compound eye.

On the other hand, it must be admitted that there are reasons for considering the different kinds of eyes to be of perfectly distinct origin. The eye of *Limulus*, according to Grenacher, is formed on a plan quite unlike that of other Crustacea. Again, the development of the eye in *Musca*, to judge from Weismann's observations, is very dissimilar from that of other insects. The varied position of the eye in different groups, as, for instance, in *Pecten*, *Spondylus*, *Euphausia*, *Onchidium*, &c., point to the same conclusion.

It seems clear that the image produced by the ocelli must be altogether different from the picture given by the compound eyes; and we may therefore reasonably conclude that the two organs have distinct functions. It used formerly to be supposed that the compound eyes were intended for distant, the ocelli for near vision. Claparède, however, has maintained the opposite theory, while Mr. Lowne regards the ocelli as incapable of producing 'anything worthy the name of an image,' and suspects that their function 'is the perception of the intensity in the direction of light, rather than vision.'

The ocelli, or simple eyes, probably see in the same manner as ours do. That is to say, the lens throws an image on the back of the eye, which we call the retina. In that case they would see everything really reversed, as we do; though long practice has given us the right impression. The simple eye of insects thus resembles ours in this respect.

As regards the mode of vision of the compound eyes, there are two distinct theories. According to one—the mosaic theory of Müller—each facet takes in only a small portion of the field; while according to the other, each facet acts as a separate eye.

This latter view has been maintained by many high authorities, but it is difficult to understand how so many images could be combined into one picture. Some insects have more than 20,000 facets on each side of their head. No ants, indeed, have so many, but in some—as, for instance, in the males of *Formica pratensis*—there are not less than 1,000. The theory, moreover, presents some great anatomical difficulties. Thus, in certain cases there is no lens, and consequently there can be no image; in some it would seem that the image would be formed completely behind the eye, while in others again it would be in front of the receptive surface. Another difficulty is that any true projection of an image would in certain species be precluded by the presence of impenetrable pigment, which only leaves a minute central passage for the light-rays. Again, it is urged that even the sharpest image would be useless, from the absence of a suitably receptive surface; since the structure of the receptive surface corresponding to each facet seems to preclude it from receiving more than a single impression.

The prevailing opinion of entomologists now is that each facet receives the impression of one pencil of rays; so that, in fact, the image formed in a compound eye is a sort of mosaic.

On the other hand, this theory itself presents great difficulties. Those ants which have very few facets must have an extremely imperfect vision. Again, while the image produced on the retina of the ocellus must of course be reversed as in our own eyes; in the compound eyes, on the contrary, the vision would, on this theory, be direct. That the same animal should see some things directly, and others reversed; and yet obtain definite conceptions of the outer world, would certainly be very remarkable.

In fact, these, so far fortunate, insects realise the epigram of Plato—

> Thou lookest on the stars, my love,
>   Ah, would that I could be
> Yon starry skies, with thousand eyes
>   That I might look on thee!

But if the male of *F. pratensis* sees 1,000 queens at once, when only one is really present, this would seem to be a bewildering privilege, and the prevailing opinion among entomologists is, as already mentioned, that each facet only takes in a portion of the object.

But while it is difficult to understand how ants see, it is clear that they do see

From the observations of Sprengel there could of

course be little, if any, doubt, that bees are capable of distinguishing colours; and I have proved experimentally that this is the case. Under these circumstances, I have been naturally anxious to ascertain, if possible, whether the same holds good with ants. I have, however, found more difficulty in doing so because, as shown in the observations just recorded, ants find their food so much more by smell than by sight.

This being so, I could not apply to ants those tests which had been used in the case of bees. At length, however, it occurred to me that I might utilize the dislike which ants, when in their nests, have to light. Of course they have no such feeling when they are out in search of food; but if light is let in upon their nests, they at once hurry about in search of the darkest corners, and there they all congregate. If, for instance, I uncovered one of my nests and then placed an opaque substance over one portion, the ants invariably collected in the shaded part.

I procured, therefore, four similar strips of glass, coloured respectively green, yellow, red, and blue, or, rather, violet. The yellow was rather paler in shade, and that glass consequently rather more transparent than the green, which, again, was rather more transparent than the red or violet. I also procured some coloured solutions.

Prof. Dewar was kind enough to test my glasses and solutions with reference to their power of transmitting colour. Taking the wave-length of the extreme visible red as 760 and that of the extreme violet as 397, we have

    760 to 647 give red.
    647 ,, 585 ,, orange.
    585 ,, 575 ,, yellow.
    575 ,, 497 ,, green.
    497 ,, 455 ,, blue.
    445 ,, 397 ,, violet.

The result of his examination of my glasses and solutions was as follows:—

The light-yellow glass cut off the high end down to wave-length 442.

The dark-yellow glass cut off the high end down to wave-length 493.

The green glass cut off the high end down to wave-length 465, and also the red to 616.

The red glass cut off the high end down to wave-length 582.

The violet glass cut off the orange and yellow from wave-length 684 to 583, and a band between wave-lengths 543 and 516.

The purple glass cut off the high end down to wave-length 528.

The solution of chromate of potash cut off the high end to 507.

The saffron cut off the high end to about 473.

The blue fluid cut off the low end to 516.

The red fluid cut off the high end to 596.

I then (July 15, 1876) laid the strips of glass on one of my nests of *Formica fusca*, containing about 170 ants. These ants, as I knew by many previous observations, seek darkness, at least when in the nest, and would collect in the darkest part. I then, after counting the ants under each strip, moved the glasses, at intervals of about half an hour, so that each should by turns cover the same portion of the nest. The results were as follows—the numbers indicating the approximate numbers of ants under each glass (there were sometimes a few not under any of the strips of glass):—

1.	Green.	Yellow.	Red.	Violet.
	50	40	80	0
2.	Violet.	Green.	Yellow.	Red.
	0	20	40	100
3.	Red.	Violet.	Green.	Yellow.
	60	0	50	50
4.	Yellow.	Red.	Violet.	Green.
	50	70	1	40
5.	Green.	Yellow.	Red.	Violet.
	30	30	100	0
6.	Violet.	Green.	Yellow.	Red.
	0	14	5	140
7.	Red.	Violet.	Green.	Yellow
	50	0	40	70
8.	Yellow.	Red.	Violet.	Green.
	40	50	1	70

9.	Green.	Yellow.	Red.	Violet.
	60	35	65	0
10.	Violet.	Green.	Yellow.	Red.
	1	50	40	70
11.	Red.	Violet.	Green.	Yellow.
	50	2	50	60
12.	Yellow.	Red.	Violet.	Green.
	35	55	0	70

Adding these numbers together, there were, in the twelve observations, under the red 890, under the green 544, under the yellow 495, and under the violet only 5. The difference between the red and the green is very striking, and would doubtless have been more so, but for the fact that when the colours were transposed the ants which had collected under the red sometimes remained quiet, as, for instance, in cases 7 and 8. Again, the difference between the green and yellow would have been still more marked but for the fact that the yellow always occupied the position last held by the red, while, on the other hand, the green had some advantage in coming next the violet. In considering the difference between the yellow and green, we must remember also that the green was decidedly more opaque than the yellow.

The case of the violet glass is more marked and more interesting. To our eyes the violet was as opaque as the red, more so than the green, and much more so than the yellow. Yet, as the numbers show, the ants had scarcely any tendency to congregate under it. There were nearly as many under the same area of the uncovered portion of the nest as under that shaded by the violet glass.

Lasius flavus also showed a marked avoidance of the violet glass.

I then experimented in the same way with a nest of *Formica fusca*, in which there were some pupæ, which were generally collected in a single heap. I used glasses coloured dark yellow, dark green, light yellow, light green, red, violet, and dark purple. The colours were always in the preceding order, but, as before, their place over the nest was changed after every observation.

To our eyes the purple was almost black, the violet and dark green very dark and quite opaque; the pupæ could be dimly seen through the red, rather more clearly through the dark yellow and light green, while the light yellow were almost transparent. There were about 50 pupæ, and the light was the ordinary diffused daylight of summer.

These observations showed a marked preference for the greens and yellows. The pupæ were $6\frac{1}{2}$ times under dark green, 3 under dark yellow, $3\frac{1}{2}$ under red, and once each under light yellow and light green, the violet and purple being altogether neglected.

I now tried the same ants under the same colours, but in the sun; and placed a shallow dish containing some 10 per cent. solution of alum sometimes over the yellow, sometimes over the red. I also put four thicknesses of violet glass, so that it looked almost black.

Under these circumstances, the pupæ were placed under the red 7 times, dark yellow 5, once they were half under each, but never under the violet, purple, light yellow, dark or light green.

The following day I placed over the same nest, in the sun, dark green glass, dark red, and dark yellow. In nine observations the pupæ were carried three times under the red and nine times under the yellow.

I then tried a similar series of experiments with *Lasius niger*, using a nest in which were about 40 pupæ, which were generally collected in a single heap all together. As before, the glasses were moved in regular order after each experiment; and I arranged them so that the violet followed the red. As far, therefore, as position was concerned, this gave violet rather the best place. The glasses used were dark violet, dark red, dark green, and yellow, the yellow being distinctly the most transparent to our eyes.

Experiment			Experiment		
1.	Pupæ under yellow.		8.	Pupæ under	green.
2.	,,	,,	9.	,,	red.
3.	,,	,,	10.	,,	yellow.
4.	,,	,,	11.	,,	red.
5.	,,	,,	12.	,,	yellow.
6.	,,	,,	13.	,,	,,
7.	,,	green.	14.	,,	red.

EXPERIMENTS WITH COLOURED GLASSES.

Experiment			Experiment		
15. Pupæ under green.			24. Pupæ under red.		
16.	,,	,,	25.	,,	yellow.
17.	,,	yellow.	26.	,,	red.
18.	,,	,,	27.	,,	,,
19.	,,	red.	28.	,,	,,
20.	,,	,,	29.	,,	,,
21.	,,	yellow.	30.	,,	yellow.
22.	,,	,,	31.	,,	red.
23.	,,	,,	32.	,,	green.

I now put two extra thicknesses of glass over the red and green.

33. Pupæ under red.			37. Pupæ under red.		
34.	,,	yellow.	38.	,,	,,
35.	,,	red.	39.	,,	yellow.
36.	,,	yellow.	40.	,,	red.

The result is very striking, and in accordance with the observations on *Formica fusca*. In 40 experiments the pupæ were carried under the yellow 19 times, under the red 16 times, and under the green 5 times only, while the violet was quite neglected. After the first twenty observations, however, I removed it.

I then tried a nest of *Cremastogaster scutellaris* with violet glass, purple glass, and red, yellow, and green solutions, formed respectively with fuchsine, bichromate of potash, and chloride of copper. The purple looked almost black, the violet very dark; the red and green, on the contrary, very transparent, and the yellow even more so. The yellow was not darker than a tincture of saffron. The latter indeed, to my eye, scarcely seemed to render the insects under them at all less apparent; while under the violet and purple I could not trace them at all. I altered the relative positions as before. The nest contained about 50 larvæ and pupæ.

I made thirteen trials, and in every case the larvæ and pupæ were brought under the yellow or the green —never once under any of the other colours.

Again, over a nest of *Formica fusca* containing about 20 pupæ I placed violet glass, purple glass, a weak solution of fuchsine (carmine), the same of chloride of copper (green), and of bichromate of potash (yellow, not darker than saffron).

I made eleven trials, and again, in every case the pupæ were brought under the yellow or the green.

I then tried a nest of *Lasius flavus* with the purple glass, violet glass, very weak bichromate of potash, and chloride of copper as before.

With this species, again, the results were the same as in the previous cases.

In all these experiments, therefore, the violet and purple light affected the ants much more strongly than the yellow and green.

It is curious that the coloured glasses appear to act on the ants (speaking roughly) as they would, or,

I should rather say, inversely as they would, on a photographic plate. It might even be alleged that the avoidance of the violet glass by the ants was due to their preferring rays transmitted by the other glasses. From the habits of these insects such an explanation would be very improbable. If, however, the preference for the other coloured glasses to the violet was due to the transmission and not to the absorption of rays—that is to say, if the ants went under the green rather than the violet because the green transmitted rays which were agreeable to the ants, and which the violet glass, on the contrary, stopped—then, if the violet was placed over the other colours, they would become as distasteful to the ants as the violet itself. On the contrary, however, whether the violet glass was placed over the others or not, the ants equally readily took shelter under them. Obviously, therefore, the ants avoid the violet glass because they dislike the rays which it transmits.

But though the ants so markedly avoided the violet glass, still, as might be expected, the violet glass certainly had some effect, because if it were put over the nest alone, the ants preferred being under it to being under the plain glass only.

I then compared the violet glass with a solution of ammonio-sulphate of copper, which is very similar in colour, though perhaps a little more violet, and arranged the depth of the fluid so as to make it as nearly as possible of the same depth of colour as the glass.

Approx. number of Ants under the	Exp. 1.	Exp. 2.	Exp. 3.	Exp. 4.	Exp. 5.	Exp. 6.
Glass ...	0	0	0	2	0	2
Solution...	40	80	100	80	50	70

	Exp. 7.	Exp. 8.	Exp. 9.	Exp. 10.		Total
Glass ...	0	2	3	0	...	9
Solution...	60	40	90	100	...	710

In another experiment with *Lasius niger* I used the dark yellow glass, dark violet glass, and a violet solution of 5 per cent. ammonio-sulphate of copper, diluted so as to be, to my eye, of exactly the same tint as the violet glass; in 8 observations the pupæ were three times under the violet solution, and 5 times under the yellow glass. I then removed the yellow glass, and in 10 more observations the pupæ were always brought under the solution.

It is interesting that the glass and the solution should affect the ants so differently, because to my eye the two were almost identical in colour. The glass, however, was more transparent than the solution.

To see whether there would be the same difference between red glass and red solution as between violet glass and violet solution, I then (Aug. 21) put over a nest of *Formica fusca* a red glass and a solution of carmine, as nearly as I could make it of the same tint. In 10 experiments, however, the ants were, generally speaking, some under the solution and some under the glass, in, moreover, as nearly as possible equal numbers.

August 20.—Over a nest of *Formica fusca* containing 20 pupæ, I placed a saturated solution of bichromate of potash, a deep solution of carmine, which let through scarcely any but the red rays, and a white porcelain plate.

Obs.
1. Under the bichr. of potash were 0 pupæ, carmine 18, porcelain 2

Obs.							
2.	„	„	0	„	„	6	„	14
3.	„	„	6	„	„	3	„	11
4.	„	„	0	„	„	5	„	18
5.	„	„	6	„	„	4	„	10
6.	„	„	0	„	„	19	„	1
7.	„	„	0	„	„	0	„	20
8	„	„	4	„	„	15	„	1
9.	„	„	2	„	„	4	„	14
10.	„	„	0	„	„	4	„	16
11	„	„	0	„	„	3	„	17
Total			18			81		124

I then put over another nest of *Formica fusca* four layers of red glass (which, when examined with the spectroscope, let through red light only), four layers of green glass (which, examined in the same way, transmitted nothing but a very little green), and a porcelain plate. Under these circumstances the ants showed no marked preference, but appeared to feel equally protected, whether they were under the red glass, the green glass, or the porcelain.

Thus, though it appears from other experiments that ants are affected by red light, still the quantity that passes through dark red glass does not seem greatly to disturb them. I tested this again by placing over a nest containing a queen and about 10 pupæ a piece of

EXPERIMENTS ON A QUEEN ANT. 197

opaque porcelain, one of violet, and one of red glass, al of the same size. The result is shown below.

Obs.						
1. Queen went under red glass			5 { pupæ were taken under red glass	2 { under porcelain
2.	,,	porcelain	0	,,	7	,,
3.	,,	red glass	0	,,	7	,,
4.	,,	,,	6	,,	2	,,
5.	,,	,,	6	,,	2	,,
6.	,,	,,	3	,,	7	,,
7.	,,	,,	10	,,	0	,,
8.	,,	,,	4	,,	6	,,
9.	,,	,,	1	,,	0	,,
10.	,,	porcelain	0	,,	10	,,
11.	,,	red glass	10	,,	0	,,
12.	,,	porcelain	4	,,	6	,,
13.	,,	red glass	7	,,	3	,,
14.	,,	porcelain	4	,,	6	,,
15.	,,	red glass	4	,,	6	,,
16.	,,	porcelain	0	,,	10	,,
17.	,,	red glass	10	,,	0	,,
18.	,,	,,	8	,,	2	,,
19.	,,	porcelain	7	,,	3	,,
20.	,,	,,	1	,,	9	,,
	Total	... ...	90		88	

Obviously, therefore, the ants showed no marked preference for the porcelain. On one, but only on one occasion (Obs. 9), most of the pupæ were carried under the violet glass, but generally it was quite neglected.

I now tried a similar experiment with porcelain and yellow glass.

Obs.						
1. Queen went under porcelain			8 { pupæ were taken under yellow.	2 { under porcelain
2.	,,	,,	2	,,	8	,,
3.	,,	,,	8	,,	2	,,

Obs.

4. Queen went under yellow glass	5	{pupæ were taken under yellow	5	{under porcelain		
5.	,,	porcelain	3	,,	8	,,
6.	,,	yellow glass	8	,,	3	,,
7.	,,	porcelain	6	,,	5	,,
8.	,,	,,	0	,,	7	,,
9.	,,	,,	0	,,	10	,,
10.	,,	yellow glass	5	,,	5	,,
11.	,,	porcelain	8	,,	2	,,
12.	,,	,,	3	,	7	,,
13.	,,	yellow glass	10	,,	0	,,
14.	,,	porcelain	0	,,	10	,,
15.	,,	yellow glass	10	,,	0	,,
16.	,,	,,	7	,,	3	,,
17.	,,	,,	10	,,	0	,,
18.	,,	porcelain	1	,,	9	,,
19.	,,	,,	0	,,	10	,,
			98		92	

The porcelain and yellow glass seemed, therefore, to affect the ants almost equally.

I then put two ants on a paper bridge, the ends supported by pins, the bases of which were in water. The ants wandered backwards and forwards, endeavouring to escape. I then placed the bridge in the dark and threw the spectrum on it, so that successively the red, yellow, green, blue, and violet fell on the bridge.

The ants, however, walked backwards and forwards without (perhaps from excitement) taking any notice of the colour.

I then allowed some ants (*Lasius niger*) to find some larvæ, to which they obtained access over a narrow paper bridge. When they had got used to it,

I arranged so that it passed through a dark box, and threw on it the principal colours of the spectrum, namely, red, yellow, green, blue, and violet, as well as the ultra-red and ultra-violet; but the ants took no notice.

It is obvious that these facts suggest a number of interesting inferences. I must, however, repeat the observations and make others; but we may at least, I think, conclude from the preceding that:—(1) ants have the power of distinguishing colours; (2) that they are very sensitive to violet; and it would also seem (3) that their sensations of colour must be very different from those produced upon us.

But I was anxious to go beyond this, and to attempt to determine how far their limits of vision are the same as ours. We all know that if a ray of white light is passed through a prism, it is broken up into a beautiful band of colours—the spectrum. To our eyes this spectrum is bounded by red at the one end and violet at the other, the edge being sharply marked at the red end, but less abruptly at the violet. But a ray of light contains, besides the rays visible to our eyes, others which are called, though not with absolute correctness, heat-rays and chemical rays. These, so far from falling within the limits of our vision, extend far beyond it, the heat-rays at the red, the chemical rays at the violet end.

I have tried various experiments with spectra derived from sunlight; but, owing to the rotation of the earth, they were not thoroughly satisfactory. Mr. Spottiswoode was also good enough to enable me to make some experiments with electric light, which were not very conclusive; more recently I have made some additional and much more complete experiments, through the kindness of Prof. Dewar, Prof. Tyndall, and the Board of Managers of the Royal Institution, to whom I beg to offer my cordial thanks.

Of course, the space occupied by the visible spectrum is well marked off by the different colours. Beyond the visible spectrum, however, we have no such convenient landmarks, and it is not enough to describe it by inches, because so much depends on the prisms used. If, however, paper steeped in thalline is placed in the ultra-violet portion of the spectrum, it gives, with rays of a certain wave-length, a distinctly visible green colour, which therefore constitutes a green band, and gives us a definite, though rough, standard of measurement.

In the above experiments with coloured spectra, the ants carried the pupæ out of the portion of the nest on which coloured light was thrown and deposited them against the wall of the nest; or, if I arranged a nest of *Formica fusca* so that it was entirely in the light, they carried them to one side or into one corner. It seemed to me, therefore, that it would be interesting so to arrange matters, that on quitting the spectrum, after passing through a dark space, the ants should encounter not a solid obstacle, but a barrier of light. With this object, I prepared some nests 12 inches long by 6 inches wide; and Mr. Cottrell kindly arranged for me at the Royal Institution on the 29th of June, by means of the electric light, two spectra, which were thrown by two glass prisms on to a table at an angle of about 45°. Each occupied about 6 inches square, and there was a space of about 2 inches between them—that is, between the red end of the one and the violet of the other.

Experiment 1.—In one of the spectra I placed a nest of *Formica fusca*, 12 inches by 6, containing about 150 pupæ, and arranged it so that one end was distinctly beyond the limit of the violet visible to us, and all but to the edge of the green given by thalline paper, and the other just beyond the visible red. The pupæ at first were almost all in or beyond the violet, but were carried into the dark space between the two spectra, the bright thalline band being avoided, but some pupæ being deposited in the red.

Experiment 2.—I then tried the same experiment with a nest of *Lasius niger*, in which there were many larvæ as well as pupæ. They were all at the commencement at the blue end of the nearer spectrum. The larvæ were left by themselves in the violet, while pupæ were ranged from the end of the green to that of the red inclusive.

Experiment 3.—Arranged a nest of *L. niger* as before; at the commencement the pupæ and larvæ were much scattered, being, however, less numerous in the violet and ultra-violet rays. Those in the ultra-violet rays were moved first, and were deposited, the larvæ in the violet, and the pupæ in the red.

Experiment 4.—Made the same experiment with another nest of *L. niger*. At the commencement the larvæ and pupæ were in the violet and ultra-violet portion, extending to double the distance from the visible end to the thalline band. The ants soon began bringing the pupæ to the red. Over part of the red I placed a piece of money. The pupæ were cleared from the ultra-violet first. That the pupæ were not put in the red for the sake of the red light was evident, because the space under the coin was even more crowded than the rest. The pupæ were heaped up in the dark as far as the thalline band of the other spectrum. I then brought the second spectrum nearer to the first. The pupæ which thus came to be in the thalline band were gradually moved into the dark.

Experiment 5.—Tried the same with another nest of *L. niger*. The pupæ were at first in the violet and ultra-violet about double as far as the thalline line, while most of the larvæ were in the green. The furthest part was cleared first; and they were again brought principally into the yellow, red, and dark.

Again, I scattered them pretty equally, some being in the ultra-violet portion, as far as double the distance of the thalline from the violet; most, however, being in the violet and blue.

The ants began by removing the pupæ which **were**

in and near the thalline band, and carried them into the yellow or red.

Experiment 6.—Repeated the same experiment. Begun it at 11.15. Placed some pupæ in the red, some in the yellow, and a few scattered over the second spectrum; there were none in the nearer one.

They were all carried away from the red past the violet, and put down in the dark portion, or in the red and yellow, of the nearer spectrum.

These experiments surprised me much at the time, as I had expected all the pupæ to be carried into the space between the two spectra; but it afterwards occurred to me that the ultra-violet rays probably extended further than I had supposed, so that even the part which lay beyond the thalline band contained enough rays to appear light to the ants. Hence perhaps they selected the red and yellow as a lesser evil.

Experiment 7.—I altered, therefore, the arrangement. Prof. Dewar kindly prepared for me a condensed pure spectrum (showing the metallic lines) with a Siemens' machine, using glass lenses and a mirror to give a perpendicular incidence when thrown on the nest. I arranged the pupæ again in the ultra-violet as far as the edge of the fluorescent light shown with thalline paper. The pupæ were all again removed, and most of them placed just beyond the red, but none in the red or yellow.

Experiment 8.—Arranged the light as before, and placed the pupæ in the ultra-violet rays. In half an hour they were all cleared away and carried into the dark space beyond the red. We then turned the nest round so that the part occupied by the pupæ again came to be in the violet and ultra-violet. The light chanced to be so arranged that along one side of the nest was a line of shadow; and into this the pupæ were carried, all those in the ultra-violet being moved. We then shifted the nest a little, so that the violet and ultra-violet fell on some of the pupæ. These were then all carried into the dark, the ones in the ultra-violet being moved first.

In these experiments with the vertical incidence there was less diffused light, and the pupæ were in no case carried into the red or yellow.

Experiment 9.—I arranged the light and the ants as before, placing the pupæ in the ultra-violet, some being distinctly beyond the bright thalline band. The ants at once began to remove them. At first many were deposited in the violet, some, however, being at once carried into the dark beyond the red. When all had been removed from the ultra-violet, they directed their attention to those in the violet, some being carried, as before, into the dark, some into the red and yellow. Again, when those in the violet had all been removed, they began on the pupæ in the red and yellow, and carried them also into the dark. This took nearly half an hour. As I had arranged the pupæ so that it might be said that they were awkwardly placed, we then turned the nest round, leaving the pupæ otherwise as they had been arranged by the ants; but the result of moving the nest was to bring some of them into the violet, though most were in the ultra-violet. They were, as before, all carried into the dark space beyond the red in about half an hour.

We then turned the glass round again, this time arranging the end about the length of the spectrum beyond the end of the violet visible to our eyes. They began clearing the thalline band, carrying some into the violet, but the majority away further from the spectrum. In a quarter of an hour the thalline band had been quite cleared; and in half an hour a band beyond, and equal to the thalline band, those in the violet being left untouched. After the pupæ in the ultra-violet portion had all been moved, those in the violet were also carried away and deposited about twice as far from the edge of the violet as the further edge of the bright thalline band.

Experiment 10.—Experimented again with the same arrangement as before, using another nest of *Lasius niger* and placing the pupæ in the violet and a little beyond. The ants at once began removing them into the dark, tunnelling into the heap, and then carrying away those in the ultra-violet first, although they were further off. In half an hour they had all been moved out of the violet and ultra-violet, about half being placed in the dark, and half having been provisionally deposited in the red and yellow.

Experiment 11.—Same arrangement as before. The pupæ being placed all along one side of the nest, from the edge of the red to a distance beyond the violet as great as the whole length of the spectrum. I began at 4.15. By degrees they were all cleared away from the spectrum, except those in the violet, where indeed, and immediately outside of which, the others were placed. At 5, however, they began to carry them back into the red. At 5.45 the blue and violet were nearly cleared, the pupæ being placed in the red and yellow. At 6.15 they had all been brought from the violet and ultra-violet into the red and yellow.

I then shook up the pupæ so that they were arranged all along one side of the nest, and extended about an inch beyond the red. This excited the ants very much, and in less than ten minutes all those in the spectrum, and for about 6 inches beyond the violet, were moved, but at first they were put down anywhere, so that they were scattered all over the nest. This, however, lasted for a very short time, and they were all carried into the dark beyond the red, or into the extreme end at some distance beyond the violet. At 7 the edge of the heap of pupæ followed the line of the red at one end, coming about $\frac{1}{4}$ inch within it, which was not owing to want of room, as one side of the nest was almost unoccupied; at the other end they were all carried 3 inches beyond the end of the violet.

It would seem, then, as the result of these experiments, that the limits of vision of ants at the red end of the spectrum are approximately the same as ours, that they are not sensitive to the ultra-red rays; but, on the other hand, that they are very sensitive to the ultra-violet rays, which our eyes cannot perceive.

I then arranged the same ants in a wooden frame consisting of a base and two side walls, between which in the middle was a perpendicular sliding door. The pupæ had been arranged by the ants in the centre of the nest, so that some were on each side of the door. We then threw, by means of a strong induction-coil, a magnesium-spark on the nest from one side, and the light from a sodium-flame in a Bunsen burner on the other, the light being in each case stopped by the sliding door, which was pressed close down on the nest. In this way the first half was illuminated by the one light, the second by the other, the apparatus being so arranged that the lights were equal to our eyes—that, however, given by the magnesium, consisting mainly of blue, violet, and ultra-violet rays, that of the sodium being very yellow and poor in chemical rays. In a quarter of an hour the pupæ were all carried into the yellow. The sodium light being the hotter of the two, to eliminate the action of heat I introduced a water-cell between the ants and the sodium-flame, and made the two sides as nearly as possible equally light to my eye. The pupæ, however, were again carried into the sodium side.

I repeated the same experiment as before, getting the magnesium-spark and the sodium-flame to the same degree of intensity, as nearly as my eye could judge, and interposing a water-screen between the sodium-flame and the ants. The temperature was tested by the thermometer, and I could distinguish no difference between the two sides. Still the ants preferred the sodium side. This I repeated twice. I then removed the magnesium-spark somewhat, so that the illumination on that side was very much fainter than on the other; still the pupæ were carried into the sodium-light. I then turned the nest round so as to bring them back into the magnesium. They were again carried to the sodium side.

Once more I repeated the same experiment. The light on the magnesium side was so faint that I could scarcely see the pupæ, those on the sodium side being quite plain. The thermometer showed no difference between the two sides. The pupæ were carried into the sodium-light. I then turned the nest round twice; but the pupæ were each time carried out of the magnesium-light.

These experiments seemed strongly to indicate, if not to prove, that ants were really sensitive to the ultra-violet rays. Now to these rays sulphate of quinine and bisulphide of carbon are extremely opaque, though perfectly transparent in the case of visible rays, and therefore to our eyes entirely colourless and transparent. If, therefore, the ants were really affected by the ultra-violet rays, then a cell containing a layer of sulphate of quinine or bisulphide of carbon would tend to darken the underlying space to their eyes, though to ours it would not do so.

It will be remembered that if an opaque substance is placed over a part of a glass nest, other things being equal, the ants always congregate under it; and that if substances of different opacity are placed on different parts of a nest, they collect under that which seems to them most opaque. Over one of my nests of *Formica fusca*, therefore, I placed two pieces of dark-violet glass 4 inches by 2 inches; and over one of them I placed a cell containing a layer of bisulphide of carbon, an inch thick, slightly coloured with iodine. In all these experiments, when I moved the liquids or glasses, I gave the advantage, if any, to the one under which experience showed that the ants were least likely to congregate. The ants all collected under the glass over which was the bisulphide of carbon.

I then thought that though no doubt the iodine rendered the bisulphide more completely impervious to the ultra-violet rays, I would try the effect of it when pure and perfectly colourless. I therefore tried the same experiment with pure bisulphide, moving the two glasses from time to time in such a manner that the ants had to pass the first violet glass in order to reach that over which was the bisulphide.

At 8.30 the ants were all under the glass over which was the bisulphide of carbon: I then changed the position.

8.45	,,	,,	,,
9	,,	,,	,,
9.15	,,	,,	,,

Although the bisulphide of carbon is so perfectly transparent, I then thought I would try it without the violet glass. I therefore covered part of the nest with violet glass, a part with a layer of bisulphide of carbon, moving them from time to time as before, and the ants in every case went under the bisulphide.

I then reduced the thickness of the layer of bisulphide to $\frac{4}{10}$ of an inch, but still they preferred the bisulphide.

Then thinking that possibly the one shelter being a plate of glass and the other a liquid might make a difference, I tried two similar bottles, one containing water and the other bisulphide of carbon; but in every case the ants went under the bisulphide of carbon. On the other hand, when I used coloured solutions so deep in tint that the ants were only just visible through them, the ants went under the coloured liquids.

October 10.--I uncovered the nest at 7 A.M., giving the ants an option between the bisulphide of carbon and various coloured solutions, taking for violet ammonio-sulphate of copper; for red, a solution of carmine so deep in tint that the ants could only just be seen through it; for green, a solution of chlorate of copper; and for yellow, saffron. They were each separately tried with the bisulphide, and in every case the ants preferred the coloured solution.

I now took successively red, yellow, and green glass; but in every case the ants preferred the glass to the bisulphide. Although, therefore, it would seem from the previous experiments that the bisulphide darkened the nests to the ants more than violet glass, it would appear to do so less than red, green, or yellow.

I now made some experiments in order, if possible, to determine whether the reason why the ants avoided the violet glass was because they disliked the colour violet, or whether it was because the violet glass transmitted more of the ultra-violet rays.

For this purpose I placed a layer of the bisulphide of carbon over a piece of violet glass. By this arrangement I got the violet without the ultra-violet rays; and I then contrasted this combination with other coloured media.

First, I took a solution of bichromate of potash (bright orange), and placed it on a part of the nest, side by side with the violet glass and bisulphide of carbon. I should add that the bichromate of potash also cuts off the ultra-violet rays. In all the following observations I changed the position after each observation.

At 1.30 P.M. the ants were under the bichromate.
3 ,, ,, half under the bichromate and half under the violet glass and bisulphide.
8 A.M. ,, ,, under the bichromate.
8.30 ,, ,, under the violet glass and bisulphide.
9 ,, ,, half under each.
9.30 ,, ,, some under each, but most

			under the violet glass and bisulphide.
9.45	,,	,,	half under each.
10	,,	,,	,, ,,

In this case, therefore, though without the layer of bisulphide the violet glass would always have been avoided, the result of placing the bisulphide over the violet glass was that the ants did not care much whether they were under the violet glass or under the bichromate of potash.

I then took the same solution of carmine which I had already used.

10.	The ants were under the carmine.
10.15	,, ,, ,,
10.30	,, most under the carmine, but some under the violet.
10.45	,, under the carmine.
11.	,, most under the carmine, but some under the violet.

Here, then, again the bisulphide made a distinct difference, though not so much so as with the bichromate of potash.

I then took the solution of chlorate of copper already used.

1	About half the ants were under each.
1.30.	The greater number were under the violet glass and bisulphide.

2. The greater number were under the violet glass and bisulphide.
2.30 „ „ „
3. Almost all were under the glass and bisulphide.

The addition of the bisulphide thus caused the violet glass to be distinctly preferred to the chlorate of copper.

I then took a solution of sulphate of nickel, almost exactly the same tint as, or a shade paler than, the chlorate of copper.

At 3.45 the ants were under the violet glass and bisulphide.
4. „ „ „
5. „ „ „

October 18.
7 A.M. „ „ „
8. About half of the ants were under each.

Here the effect was even more marked.

I then took some saffron 1 inch in thickness, and of a deep-yellow colour.

12.45 The ants were about half under each.
1. Most of the ants were under the violet glass and bisulphide.
1.15 „ „ „
2. Most of the ants were under the saffron.

Here, again, we have the same result.

I then tried the different-coloured glasses, all of which, as I had previously found, are unmistakably preferred to the violet. It remained to be seen what effect placing the bisulphide of carbon on the violet would have.

First, I placed side by side, as usual, a piece of green glass and the violet glass covered with bisulphide of carbon: —

 1st exp. Half of the ants were under each.
 2nd ,, They were under the violet glass and bisulphide.
 3rd ,, ,, ,, ,,
 4th exp. Most of them were under the violet glass and bisulphide.
 5th ,, ,, ,, ,,

Next, I tried pale-yellow glass.

 1st obs. The ants were almost all under the violet glass and bisulphide.
 2nd ,, About three-quarters were ,, ,,
 3rd ,, They were all ,, ,,
 4th ,, About half were under each.

I then took the dark-yellow glass.

 1st obs. About half the ants were under the yellow glass and half under the violet glass and bisulphide.
 2nd ,, Most of them were under the violet glass and bisulphide.
 3rd ,, ,, ,, yellow glass.
 4th ,, ,, ,, violet glass and bisulphide.
 5th ,, About half under each.

I now took deep-red glass.

    1st obs. The ants were under the red glass.
    2nd ,, Half of the ants were under each.
    3rd ,, Most of the ants were under violet glass and bisulphide.
    4th ,, Half were under each.

It seemed evident, therefore, that while if violet glass alone was placed side by side with red, yellow, or green, the ants greatly preferred any of the latter, on the other hand, if a layer of bisulphide of carbon, which to our eyes is perfectly transparent, was placed over the violet glass, they then went as readily, or even more readily, under it than under other colours.

In order to be sure that it was not the mere presence of a fluid, or the two layers of glass, to which this was due, I thought it would be well to try a similar series of experiments, using, however, a layer of similar thickness (1 inch) of water coloured light blue by ammonio-sulphate of copper.

I therefore took again the piece of violet glass, over which I placed a flat-sided bottle, about 1 inch thick, containing a light-blue solution of ammonio-sulphate of copper; and, in contrast with it, I used the same coloured glasses as before. The difference, however, was very marked, the ants always preferring the red, green, and yellow to the violet.

These experiments seem to demonstrate that in the previous series the ants were really influenced by some difference due to the bisulphide of carbon, which affected their eyes, though not ours.

I then thought it would be interesting to use, instead of the bisulphide, a solution of sulphate of quinine ($\frac{1}{2}$ dr. to 4 ounces), which differs from it in many points, but agrees in cutting off the ultra-violet rays. I used, as before, a layer about an inch thick, which I placed over violet glass, and then placed by its side the same coloured glasses as before.

First, I took the red glass.

Obs. 1. About half the ants were under each.
 „ 2. Most of them were under the red glass.
 „ 3. About half under each; rather more under the violet glass and sulphate of quinine than under the red glass.
 „ 4.   „   „   „

I now took the dark-yellow glass instead of the red.

Obs. 1. Most of the ants were under the violet glass and sulphate of quinine.
 „ 2. All   „   „   „
 „ 3. „   „   „   „
 „ 4. „   „   „   yellow glass.
 „ 5. „   „   „   „
 „ 6. All the ants were under the violet glass and sulphate of quinine.
 „ 7. About half under each.
 „ 8. Rather more under the violet glass and sulphate of quinine than under the yellow glass.

I then took the light-yellow glass instead of the dark.

Obs. 1. The ants were all under the violet glass and sulphate of quinine.
 „ 2. Rather more than half under tne yellow glass.
 „ 3. Almost all under the violet glass and sulphate of quinine.
 „ 4. All „ „ „ „

I then took the green glass instead of the yellow.

Obs. 1. They were under the violet glass and sulphate of quinine.
 „ 2. „ „ „
 „ 3. About half under each.
 „ 4. About three-quarters under the green glass.
 „ 5. Almost all under the violet glass and sulphate of quinine.

Thus, then, while if the ants have to choose between the violet and other coloured glasses, they will always prefer one of the latter, the effect of putting over the violet glass a layer either of sulphate of quinine or bisulphide of carbon, both of which are quite transparent, but both of which cut off the ultra-violet rays, is to make the violet glass seem to the ants as good a shelter as any of the other glasses. This seems to me strong evidence that the ultra-violet rays are visible to the ants.

I then tried similar experiments with a saturated solution of chrome alum and chromium chloride. These are dark greenish blue, very opaque to the visible light-rays, but transparent to the ultra-violet. I used a layer $\frac{1}{4}$ inch thick, which was still so dark that I could not see the ants through it; and for comparison, a solution 1 inch thick of bisulphide of carbon, moving them after each observation as before.

Exp. 1. The ants were under the bisulphide of carbon.
„ 2. „ „ „ „
Exp. 3. Most of the ants were under the bisulphide of carbon.
„ 4. All but three „ „
„ 5. All „ „

I now took chromium chloride instead of alum.

Exp. 1. Most were under the bisulphide of carbon.
„ 2. All „ „ „
„ 3. Almost all „ „ „
„ 4. About three-fourths were under the chromium chloride.
„ 5. All were under the chromium chloride.
„ 6. About two-thirds „ „
„ 7. About one-half under each.
„ 8. All under the bisulphide of carbon.
„ 9. About three-fourths under the bisulphide of carbon.
„ 10. About half „ „ „
„ 11. All under the chrome alum.
„ 12. „ bisulphide of carbon.

This result is very striking. It appears to show that though to our eyes the bisulphide of carbon is absolutely transparent, while the chrome alum and chromium chloride are very dark, to the ants, on the contrary, the former appears to intercept more light than a layer of the latter, which to our eyes appears dark green.

The only experiments hitherto made with the view of determining the limits of vision of animals have been some by Prof. Paul Bert [1] on a small fresh-water crustacean belonging to the genus *Daphnia*, from which he concludes that they perceive all the colours known to us, being, however, specially sensitive to the yellow and green, and that their limits of vision are the same as ours.

Nay, he even goes further than this, and feels justified in concluding from the experience of two widely divergent species—Man and *Daphnia*—that the limits of vision would be the same in all cases.

His words are—

A. 'Tous les animaux voient les rayons spectraux que nous voyons.'

B. 'Ils ne voient aucun de ceux que nous ne voyons pas.'

C. 'Dans l'étendue de la région visible, les différences entre les pouvoirs éclairants des différents rayons coloriés sont les mêmes pour eux et pour nous.'

He adds, that 'puisque les limites de visibilités semblent être les mêmes pour les animaux et pour nous,

---

[1] *Archiv. de Physiol.* 1869, p. 547.

ne trouvons-nous pas là une raison de plus pour supposer que le rôle des milieux de l'œil est tout-à-fait secondaire, et que la visibilité tient à l'impressionnabilité de l'appareil nerveux lui-même?'

Such a generalisation would seem to rest on but a slight foundation; and I may add that I have made some experiments myself[1] on Daphnias which do not agree with those of M. Bert. On the contrary, I believe that the eyes of Daphnias are in this respect constituted like those of ants.

These experiments seem to me very interesting. They appear to prove that ants perceive the ultra-violet rays. Now, as every ray of homogeneous light which we can perceive at all appears to us as a distinct colour, it becomes probable that these ultra-violet rays must make themselves apparent to the ants as a distinct and separate colour (of which we can form no idea), but as unlike the rest as red is from yellow, or green from violet. The question also arises whether white light to these insects would differ from our white light in containing this additional colour. At any rate, as few of the colours in nature are pure, but almost all arise from the combination of rays of different wavelengths, and as in such cases the visible resultant would be composed not only of the rays which we see, but of these and the ultra-violet, it would appear that the colours of objects and the general aspect of nature

[1] *British Assoc. Report* 1881, and *Linnæan Soc. Journ.* 1882

must present to them a very different appearance from what it does to us.

The Sense of Hearing.

Many eminent observers have regarded the antennæ of insects as auditory organs, and have brought forward strong evidence in favour of their view.

I have myself made experiments on grasshoppers, which convinced me that their antennæ serve as organs of hearing.

So far, however, as Ants, Bees, and Wasps are concerned, the evidence is very conflicting. The power of hearing has indeed generally been attributed to them. Thus St. Fargeau, in his 'Hist. Nat. des Hyménoptères,'[1] thinks there can be no doubt on the subject. Bevan expresses, no doubt, the general opinion with reference to Bees, when he says that 'there is good evidence that Bees have a quick sense of hearing.'[2]

As regards Wasps, Ormerod, who studied them so lovingly, came to the same conclusion.[3]

On the other hand, both Huber[4] and Forel[5] state that ants are quite deaf. As I have already mentioned in the 'Linnæan Journal' (vols. xii. and xiii.), I have never succeeded in satisfying myself that my ants, bees, or wasps heard any of the sounds with

---

[1] Vol. i. p. 113. [2] *The Honey Bee*, p. 261.
[3] *Nat. Hist. of Wasps*, p 72. [4] *Nat. Hist. of Ants*.
[5] *Fourmis de la Suisse*, p. 121.

which I tried them. I have over and over again tested them with the loudest and shrillest noises I could make, using a penny pipe, a dog-whistle, a violin, as well as the most piercing and startling sounds I could produce with my own voice, but all without effect. At the same time, I carefully avoided inferring from this that they are really deaf, though it certainly seems that their range of hearing is very different from ours.

In order, if possible, to throw some light upon this interesting question, I made a variety of loud noises, including those produced by a complete set of tuning-forks, as near as possible to the ants mentioned in the preceding pages, while they were on their journeys to and fro between the nests and the larvæ. In these cases the ants were moving steadily and in a most business-like manner, and any start or alteration of pace would have been at once apparent. I was never able, however, to perceive that they took the slightest notice of any of these sounds. Thinking, however, that they might perhaps be too much absorbed by the idea of the larvæ to take any notice of my interruptions, I took one or two ants at random and put them on a strip of paper, the two ends of which were supported by pins with their bases in water. The ants imprisoned under these circumstances wandered slowly backwards and forwards along the paper. As they did so, I tested them in the same manner as before, but was unable to perceive that they took the slightest notice of any sound which I was able to produce. I then took a large female of *F. ligniperda*, and tethered her on a board to a pin by a delicate silk thread about 6 inches in length. After wandering about for a while, she stood still, and I then tried her in the same way; but, like the other ants, she took no notice whatever of the sounds.

It is of course possible, if not probable, that ants, even if deaf to sounds which we hear, may hear others to which we are deaf.

Having failed, therefore, in hearing them or making them hear me, I endeavoured to ascertain whether they could hear one another.

To ascertain then if possible whether ants have the power of summoning one another by sound, I tried the following experiments. I put out (Sept. 1874) on the board where one of my nests of *Lasius flavus* was usually fed, six small pillars of wood about an inch and a half high, and on one of them I put some honey. A number of ants were wandering about on the board itself in search of food, and the nest itself was immediately above, and about 12 inches from, the board. I then put three ants to the honey, and when each had sufficiently fed I imprisoned her and put another; thus always keeping three ants at the honey, but not allowing them to go home. If then they could summon their friends by sound, there ought soon to be many ants at the honey. The results were as follow:

September 8.—Began at 11 A.M. Up to 3 o'clock only seven ants found their way to the honey, while about as many ran up the other pillars. The arrival of these seven, therefore, was not more than would naturally result from the numbers running about close by. At 3 we allowed the ants then on the honey to return home. The result was that from 3.6, when the first went home, to 3.30, eleven came; from 3.30 to 4, no less than forty-three. Thus in four hours only seven came, while it was obvious that many would have wished to come, if they had known about the honey, because in the next three quarters of an hour, when they were informed of it, fifty-four came.

On September 10 I tried the same again, keeping as before three ants always on the honey, but not allowing any to go home. From 12 to 5.30, only eight came. Those on the honey were then allowed to take the news home. From 5.30 to 6, four came; from 6 to 6.30, four; from 6.30 to 7, eight; from 7.30 to 8, no less than fifty-one.

On September 23 we did the same again, beginning at 11.15. Up to 3.45 nine came. The ants on the honey were then allowed to go home. From 4 to 4.30 nine came; from 4.30 to 5, fifteen; from 5 to 5.30 nineteen; from 5.30 to 6, thirty-eight. Thus in three and a half hours only nine came; in two, when the ants were permitted to return, eighty-one.

Again, on September 30 I tried the same arrangement, again beginning at 11. Up to 3.30 seven ants came. We then allowed the ants which had fed to go home. From 3.30 to 4.30 twenty-eight came. From 4.30 to 5, fifty-one came. Thus in four hours and a half only seven came; while when the ants were allowed to return no less than seventy-nine came in an hour and a half. It seems obvious therefore that in these cases no communication was transmitted by sound.

Again, Professor Tyndall was good enough to arrange for me one of his sensitive flames; but I could not perceive that it responded in any way to my ants. The experiment was not, however, very satisfactory, as I was not able to try the flame with a very active nest. Professor Bell most kindly set up for me an extremely sensitive microphone: it was attached to the underside of one of my nests; and though we could distinctly hear the ants walking about, we could not distinguish any other sound.

It is, however, far from improbable that ants may produce sounds entirely beyond our range of hearing. Indeed, it is not impossible that insects may possess senses, or sensations, of which we can no more form an idea than we should have been able to conceive red or green if the human race had been blind. The human ear is sensitive to vibrations reaching at the outside to 38,000 in a second. The sensation of red is produced when 470 millions of millions of vibrations enter the eye in a similar time; but between these two numbers, vibrations produce on us only the sensation of heat;

we have no special organs of sense adapted to them. There is, however, no reason in the nature of things why this should be the case with other animals; and the problematical organs possessed by many of the lower forms may have relation to sensations which we do not perceive. If any apparatus could be devised by which the number of vibrations produced by any given cause could be lowered so as to be brought within the range of our ears, it is probable that the result would be most interesting.

Moreover, there are not wanting observations which certainly seem to indicate that ants possess some sense of hearing.

I am, for instance, indebted to Mr. Francis Galton for the following quotation from Colonel Long's recent work on Central Africa.[1] 'I observed,' he says, 'the manner of catching them' (the ants, for food), 'as here pictured' (he gives a figure). 'Seated round an ant-hole were two very pretty maidens, who with sticks beat upon an inverted gourd, "bourmah," in cadenced time to a not unmusical song, that seduced from its hole the unwary ant, who, approaching the orifice, was quickly seized.' The species of ant is not mentioned.

Moreover, there are in the antennæ certain remarkable structures, which may very probably be auditory organs.

These curious organs (Fig. 6) were first noticed,

---

[1] *Central Africa*, by Col. C. C. Long, p. 274.

so far as I am aware, by Dr. J. Braxton Hicks in his excellent paper on the 'Antennæ of Insects,' published in the 22nd volume of the 'Linnæan Transactions;' and, again, by Dr. Forel in his 'Fourmis de la Suisse.' They certainly deserve more attention than they have yet received. The cork-shaped organs (Figs. 6 and 7, $e\,e$) occur in allied species; but these stethoscope-like organs have not, so far as I am aware, been yet observed in other insects. They consist of an outer sac (Figs. 6 and 7, s), of a long tube (t), and a posterior chamber (w), to which is given a nerve (n).

Forel[1] also describes these curious organs. He appears to consider that the number varies consider-

Fig. 6.

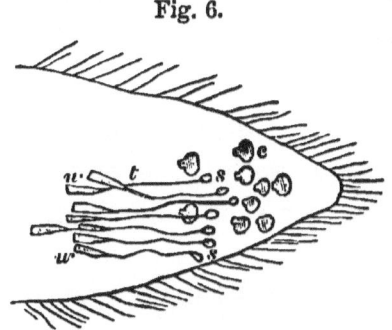

Terminal portion of antennæ of *Myrmica ruginodis* ⚥ × 75.

ably, namely, from 5 to 12. My own impression is that this difference is only apparent, and that in reality the numbers in each species vary little. Though

[1] *Trans. of Linnæan Soc.*, vol. xxii. p. 391.
[2] *Fourmis de la Suisse*, p. 301.

sometimes the presence of air renders them very conspicuous, they are in others by no means easy to make

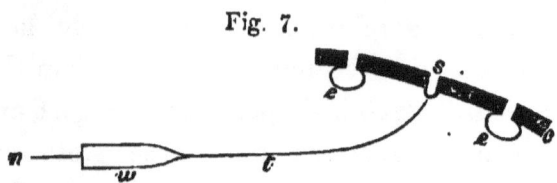

Fig. 7.

Diagrammatic section through part of Fig. 6.

c, chitinous skin of the antenna. *e e*, two of the cork-shaped organs. *s*, external chamber of one of the stethoscope-shaped organs. *t*, the tube. *w*, the posterior sac. *n*, the nerve.

out; and I think that when a small number only are apparently present, this is probably due merely to the fact that the others are not brought out by the mode of preparation.

In addition to the group of these organs situated in the terminal segment, there is one, or in some rare cases I have found two, in each of the small preceding segments. The tubes in these segments appeared to the eye to be nearly of the same length as those in the terminal segment, but I could not measure their exact length, as they do not lie flat. In some cases, when the segment was short, the tube was bent—an indication, perhaps, that the exact length is of importance. It is possible that these curious organs may be auditory, and serve like microscopic stethoscopes. Professor Tyndall, who was good enough to examine them with me, concurred in the opinion that this was very probable. I believe I am correct in saying that the bending of the tube in the short segments would make little difference in its mode of action.

Kirby and Spence were, I believe, the first to notice that an insect allied to the ants (*Mutilla Europœa*) has the power of making a sibilant, chirping sound, but they did not ascertain how this was effected. Goureau[1] subsequently called attention to the same fact, and attributed it to friction of the base of the third segment of the abdomen against the second. Westwood,[2] on the other hand, thought the sound was produced ' by the action of the large collar against the front of the mesothorax.' Darwin, in his ' Descent of Man,' adopts the same view. ' I find,' he says,[3] ' that these surfaces (*i.e.* the overlapping portions of the second and third abdominal segments) are marked with very fine concentric ridges, but so is the projecting thoracic collar, on which the head articulates; and this collar, when scratched with the point of a needle, emits the proper sound.' Landois, after referring to this opinion, expresses himself strongly in opposition to it. The true organ of sound is, he maintains,[4] a triangular field on the upper surface of the fourth abdominal ring, which is finely ribbed, and which, when rubbed, emits a stridulating sound. It certainly would appear, from Landois' observations, that this structure does produce sound, whether or not

[1] *Ann. de la Soc. Ent. de France*, 1837.
[2] *Modern Classifications of Insects*, vol. ii.
[3] *Descent of Man*, vol. i. **p.** 366.
[4] *Thierstimmen*, p. 132.

we consider that the friction of the collar against the mesothorax may also assist in doing so.

Under these circumstances, Landois asked himself whether other genera allied to Mutilla might not possess a similar organ, and also have the power of producing sound. He first examined the genus Ponera, which, in the structure of its abdomen, nearly resembles Mutilla, and here also he found a fully developed stridulating apparatus.

He then turned to the true ants, and here also he found a similar rasp-like organ in the same situation. It is indeed true that ants produce no sounds which are audible by us; still, when we find that certain allied insects do produce sounds appreciable to us by rubbing the abdominal segments one over the other; and when we find, in some ants, a nearly similar structure, it certainly seems not unreasonable to conclude that these latter also do produce sounds, even though we cannot hear them. Landois describes

Fig. 8.

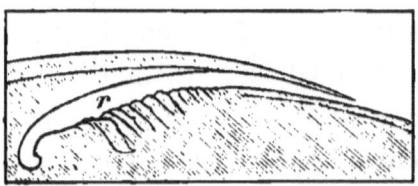

Attachment of abdominal segments of *Lasius flavus* ☿ × 225.

the structure in the workers of *Lasius fuliginosus* as having 20 ribs in a breadth of 0·13 of a millimetre,

ON ABDOMINAL SEGMENTS.

but he gives no figure. In Fig. 8 I have represented the junction of the second and third abdominal segments in *Lasius flavus*, × 225, as shown in a longitudinal and vertical section. There are about ten well-marked ribs (*r*), occupying a length of approximately $\frac{1}{100}$ of an inch. Similar ridges also occur between the following segments.

In connection with the sense of hearing I may mention another very interesting structure. In the year 1844, Von Siebold described[1] a remarkable organ which he had discovered in the tibiæ of the front legs of *Gryllus*, and which he considered to serve for the purpose of hearing. These organs have also been studied by Burmeister, Brunner, Hensen, Leydig, and others, and have recently been the subject of a monograph by Dr. V. Graber,[2] who commences his memoir by observing that they are organs of an entirely unique character,

Fig. 9.

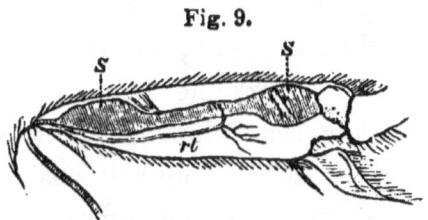

Tibia of *Lasius flavus* ⚥ × 75.

and that nothing corresponding to them occurs in any other insects, or indeed in any other Arthropods.

[1] See also Sharp, *Trans. Ent. Soc.*, 1893.

[2] *Ueber das Stimm. und Gehörorgan der Orthopteren*, Wiegmann's *Art. f. Natur.*, 1844.

[3] *Die Tympanalen Sinnesapparate der Orthopteren*, von Dr. Vitus Graber, 1875.

I have therefore been very much interested by discovering (1875) in ants a structure which seems in some remarkable points to resemble that of the Orthoptera. As will be seen from a glance at Dr. Graber's memoir, and the plates which accompany it, the large trachea of the leg in the Orthoptera is considerably swollen in the tibia, and sends off, shortly after entering the tibia, a branch which, after running for some time parallel to the principal trunk, joins it again. See, for instance, in his monograph, plate ii., fig. 43; plate vi., fig. 69; plate vii., fig. 77; &c.

Now, I have observed that in many other insects the tracheæ of the tibia are dilated, and in several I have been able to detect a recurrent branch. The same is also the case in some mites. I will, however, reserve what I have to say on this subject, with reference to other insects, for another occasion, and will at present confine myself to the ants. If we examine the tibia, say of *Lasius flavus*, Fig. 9, we shall see that the trachea presents a remarkable arrangement, which at once reminds us of that which occurs in Gryllus and other Orthoptera. In the femur it has a diameter of about $\frac{1}{3000}$ of an inch; as soon, however, as it enters the tibia, it swells to a diameter of about $\frac{1}{500}$ of an inch, then contracts again to $\frac{1}{800}$, and then again, at the apical extremity of the tibia, once more expands to $\frac{1}{500}$. Moreover, as in *Gryllus*, so also in *Formica*, a small branch rises from the upper sac, runs almost straight down the tibia, and falls again into the main trachea just above the lower sac.

The remarkable sacs (Fig. 9, *s s*) at the two extremities of the trachea in the tibia may also be well seen in other transparent species, such, for instance, as *Myrmica ruginodis* and *Pheidole megacephala*.

At the place where the upper tracheal sac contracts (Fig. 9), there is, moreover, a conical striated organ (x), which is situated at the back of the leg, just at the apical end of the upper tracheal sac. The broad base lies against the external wall of the leg, and the fibres converge inwards. In some cases I thought I could perceive indications of bright rods, but I was never able to make them out very clearly. This also reminds us of a curious structure which is found in the tibiæ of Locustidæ, between the trachea, the nerve, and the outer wall, and which is well shown in some of Dr. Graber's figures.

On the whole, then, though the subject is still involved in doubt, I am disposed to think that ants perceive sounds which we cannot hear

The Sense of Smell.

I have also made a number of experiments on the power of smell possessed by ants. I dipped camel's-hair brushes into peppermint-water, essence of cloves, lavender-water, and other strong scents, and suspended them about ¼ of an inch above the strips of paper along which the ants were passing, in the experiments above recorded. Under these circumstances, while some of the ants passed on without taking any notice, others stopped when they came close to the pencil, and, evidently perceiving the smell, turned back. Soon, however, they returned and passed the scented pencil. After doing this two or three times, they generally took no further notice of the scent. This experiment left no doubt on my mind; still, to make the matter even more clear, I experimented with ants placed on an isolated strip of paper. Over the paper, and at such a distance as almost, but not quite, to touch any ant which passed under it, I again suspended a camel's-hair brush, dipped in assafœtida, lavender-water, peppermint-water, essence of cloves, and other scents. In this experiment the results were very marked; and no one who watched the behaviour of the ants under these circumstances could have the slightest doubt as to their power of smell.

I then took a large female of *F. ligniperda* and tethered her on a board by a thread as before. When she was quite quiet I tried her with the tuning-forks; but they did not disturb her in the least. I then approached the feather of a pen very quietly, so as almost to touch first one and then the other of the antennæ, which, however, did not move. I then dipped the pen in essence of musk and did the same; the antenna was slowly retracted and drawn quite back. I then repeated the same with the other antenna. If I touched the antenna, the ant started away, apparently smarting. I repeated the same with essence of lavender, and with a second ant. The result was the same.

Many of my other experiments—for instance, some of those recorded in the next chapter—point to the same conclusion; and, in fact, there can be no doubt whatever that in ants the sense of smell is highly developed.

CHAPTER IX.

GENERAL INTELLIGENCE, AND POWER OF FINDING THEIR WAY.

A NUMBER of interesting anecdotes are on record as to the ingenuity displayed by ants under certain circumstances.

M. Lund, for instance, tells the following story as bearing on the intelligence of ants:[1]—

'Passant un jour près d'un arbre presque isolé, je fus surpris d'entendre, par un temps calme, des feuilles qui tombaient comme de la pluie. Ce qui augmenta mon étonnement, c'est que les feuilles détachées avaient leur couleur naturelle, et que l'arbre semblait jouir de toute sa vigueur. Je m'approchai pour trouver l'explication de ce phénomène, et je vis qu'à peu près sur chaque pétiole était postée une fourmi qui travaillait de toute sa force; le pétiole était bientôt coupé et la feuille tombait par terre. Une autre scène se passait au pied de l'arbre: la terre était couverte de fourmis occupées à découper les feuilles à mesure qu'elles tombaient, et les morceaux étaient sur le champ transportés dans le nid. En moins d'une heure le grand œuvre

[1] *Ann. des Sci. Nat.* 1831, p. 112.

s'accomplit sous mes yeux, et l'arbre resta entièrement dépouillé.'

Bates[1] gives an apparently similar, but really very different account. 'The Saüba ants,' he says, ' mount the tree in multitudes, the individuals being all worker-minors. Each one places itself on the surface of a leaf, and cuts with its sharp scissor-like jaws a nearly semicircular incision on the upper side; it then takes the edge between its jaws, and by a sharp jerk detaches the piece. Sometimes they let the leaf drop to the ground, where a little heap accumulates, until carried off by another relay of workers; but, generally, each marches off with the piece it has operated upon.'

Dr. Kerner recounts[2] the following story communicated to him by Dr. Gredler of Botzen :—

'One of his colleagues at Innsbrück, says that gentleman, had for months been in the habit of sprinkling pounded sugar on the sill of his window, for a train of ants, which passed in constant procession from the garden to the window. One day, he took it into his head to put the pounded sugar into a vessel, which he fastened with a string to the transom of the window; and, in order that his long-petted insects might have information of the supply suspended above, a number of the same set of ants were placed with the sugar in the vessel. These busy creatures forthwith

---

[1] *Naturalist on the Amazons*, vol. i. p. 26.
[2] *Flowers and their Unbidden Guests*, Dr. A. Kerner. Trans. by W. Ogle, 1878, p. 21.

seized on the particles of sugar, and soon discovering the only way open to them, viz. up the string, over the transom and down the window-frame, rejoined their fellows on the sill, whence they could resume the old route down the steep wall into the garden. Before long the route over the new track from the sill to the sugar, by the window-frame, transom, and string was completely established; and so passed a day or two without anything new. Then one morning it was noticed that the ants were stopping at their old place, that is, the window-sill, and getting sugar there. Not a single individual any longer traversed the path that led thence to the sugar above. This was not because the store above had been exhausted; but because some dozen little fellows were working away vigorously and incessantly up aloft in the vessel, dragging the sugar crumbs to its edge, and throwing them down to their comrades below on the sill, a sill which with their limited range of vision they could not possibly see!'

Leuckart also made a similar experiment. Round a tree which was frequented by ants, he spread a band soaked in tobacco water. The ants above the band after awhile let themselves drop to the ground, but the ascending ants were long baffled. At length he saw them coming back, each with a pellet of earth in its mouth, and thus they constructed a road for themselves, over which they streamed up the tree.

Dr. Büchner records the following instance on the authority of a friend (M. Theuerkauf):—

'A maple tree standing on the ground of the manufacturer, Vollbaum, of Elbing (now of Dantzic) swarmed with aphides and ants. In order to check the mischief, the proprietor smeared about a foot width of the ground round the tree with tar. The first ants who wanted to cross naturally stuck fast. But what did the next? They turned back to the tree and carried down aphides, which they stuck down on the tar one after another until they had made a bridge, over which they could cross the tar-ring without danger. The above-named merchant, Vollbaum, is the guarantor of this story, which I received from his own mouth on the very spot whereat it occurred.'[1]

In this case I confess I have my doubts as to the interpretation of the fact. Is it not possible that as the ants descended the tree, carrying the aphides, the latter naturally stuck to the tar, and would certainly be left there. In the same way I have seen hundreds of bits of earth deposited on the honey with which I fed my ants.

On one occasion Belt observed[2] a community of leaf-cutting ants (*Œcodoma*), which was in the process of moving from one nest to another. 'Between the old burrows and the new one was a steep slope. Instead of descending this with their burdens, they cast

[1] *Mind in Animals*, by Prof. Ludwig Büchner, p. 120.
[2] *Naturalist in Nicaragua*, O. Belt, p. 76.

them down on the top of the slope, whence they rolled to the bottom, where another relay of labourers picked them up and carried them to the new burrow. It was amusing to watch the ants hurrying out with bundles of food, dropping them over the slope, and rushing back immediately for more.'

With reference to these interesting statements, I tried the following experiment:—

October 15 (see Fig. 10).—At a distance of 10 inches from the door of a nest of *Lasius niger* I fixed an upright ash wand 3 feet 6 inches high (*a*), and from the top of it I suspended a second, rather shorter wand (*b*). To the lower end of this second wand, which hung just over the entrance to the nest (*c*), I fastened a flat glass cell (*d*) in which I placed a number of larvæ, and to them I put three or four specimens of *L. niger*. The drop from the glass cell to the upper part of the frame was only ½ an inch; still, though the ants reached over and showed a great anxiety to take this short cut home, they none of them faced the leap, but all went round by the sticks, a distance of nearly 7 feet. At 6 P.M. there were over 550 larvæ in the glass cell, and I reduced its distance from the upper surface of the nest to about ⅔ of an inch, so that the ants could even touch the glass with their antennæ, but could not reach up nor step down. Still, though the drop was so small, they all went round. At 11 P.M. the greater number of the larvæ had been carried off; so I put a fresh lot in the cell. The ants were busily at work. At 3 A.M. I visited them again. They were still carrying off the larvæ, and all going round. At 6 A.M. the larvæ were all removed. I put a fresh lot, and up to 9 A.M. they went on as before.

The following day (October 17) I took two longer sticks, each 6 feet 6 inches in length, and arranged them in a similar manner, only horizontally instead of vertically. I also placed fine earth under the glass supporting the larvæ. At 8 o'clock I placed an ant on the larvæ; she took one, and I then coaxed her home along the sticks. She deposited her larva and immediately came out again, not, however, going along the stick, but under the larvæ, vainly reaching up and endeavouring to reach the glass. At 8.30 I put her on the larvæ again, and as she evidently did not know her way home, but kept stretching herself down and trying to reach the earth under the glass cell, I again coaxed her home along the sticks. At 9.3 she came out again, and again went under the larvæ and wandered about there. At 10 I put her on the larvæ and again helped her home. At 10.15 she came out again, and this time went to the stick, but still wanted some guidance. At 10.45 she again reached the frame, but immediately came out again, and I once more coaxed her round. After wandering about some time with a larva in her mouth, she dropped down at 11.14. After depositing her larva, she came out directly and went under the larvæ. I again coaxed her round, and this time also she dropped off the glass with her larva. At 12.30 she came out again, and for the last time I helped her round. After this she found her way by herself. At 12.20 another (No. 2) found her way round and returned at 12.37. For the next hour their times were as follows:—

No. 1.	No. 2.
12.46	
	12.47
12.54	12.54
	1. 0
1. 1	
1. 7	
	1. 8
1.12	
	1.14
1.19	
	1.21
1.26	
	1.28
1.32	
	1.34
1.38	
	1.41
1.45	
	1.47
1.52	
	1.54

Thus they both made 9 visits in an hour. As regards actual pace, I found they both did about 6 feet in a minute. Soon after these began, other ants came with them. It was a beautiful day, and all my ants were unusually active. At 1 P.M. I counted 10 on the sticks at once, by 1.30 over 30, and at 5 in the afternoon over 60. They went on working very hard, and forming a continuous stream till I went to bed at 11; and at 4 in the morning I found them still at work; but though they were very anxious and, especially at first, tried very hard to save themselves the trouble of going round, they did not think of jumping down, nor did they throw the larvæ over the edge.

Moreover, as I had placed some sifted mould under the glass, a minute's labour would have been sufficient to heap up one or two particles, and thus make a little mound which would have enabled them to get up and down without going round. A mound $\frac{1}{8}$ inch high would have been sufficient; but it did not occur to them to form one.

The following morning (October 18) I put out some larvæ again at 6 A.M. Some of the ants soon came; and the same scene continued till 11.30 A.M., when I left off observing.

Again, on October 22, I placed a few larvæ in a glass, which I kept continually replenished, which was suspended $\frac{1}{3}$ of an inch above the surface of the frame containing their nest, but only connected with it by tapes five feet long. I then, at 6.30, put a *L. niger* to the larvæ; she took one and tried hard to reach down, but could not do so, and would not jump; so I coaxed her round the tapes. She went into the nest, deposited her larva, and immediately came out again. I put her back on the larvæ at 7.15; she took one, and again tried hard, but ineffectually, to reach down. I therefore again coaxed her round. She went into the nest, deposited her larva, and came out again directly as before. I put her back on the larvæ at 7.35, when the same thing happened again. She got back to the nest at 7.40, and immediately came out again. This time she found her way round the string, with some help from me, and reached the larvæ at 7.50. I helped her home for the last time. The next journey she found her way without assistance, and reached the larvæ at 8.26. After this she returned as follows, viz.:—

At 8.50
9. 0
9.10
9.17
9.28

I now made the length of the journey round the tapes 10 feet. This puzzled her a little at first.

She returned as follows:—

9.41	10.35
9.55	10.44
10. 8	10.54
10.16	11. 6
10.26	11.14 with a friend

CONSTRUCTING BRIDGES. 245

I now increased the length to 16 feet, and watched her while she made thirty journeys backwards and forwards. She also brought during the time seven friends with her.

It surprised me very much that she preferred to go so far round rather than to face so short a drop.

In illustration of the same curious fact, I several times put specimens of *L. niger* on slips of glass raised only one-third of an inch from the surface of the nest. They remained sometimes three or four hours running about on the glass, and at last seemed to drop off accidentally.

Myrmica ruginodis has the same feeling. One morning, for instance, I placed one in an isolated position, but so that she could escape by dropping one-third of an inch. Nevertheless at the same hour on the following morning she was still in captivity, having remained out twenty-four hours rather than let herself down this little distance.

Again I filled a saucer (woodcut, Fig. 11, s) with water and put in it a block of wood (w), on the top of which I fastened a projecting wooden rod (B), on the end of which I placed a shallow glass cell (A) containing several hundred larvæ. From this cell I allowed a slip of paper (P) to hang down to within $\frac{3}{10}$ of an inch of the upper surface of the nest. At one side I put another block of wood (c) with a lateral projection (D) which hung over the cell containing the larvæ. I then made a connexion between D and A, so that ants could ascend C, and, passing over D, descend upon the larvæ. I then put some specimens of *Lasius niger* to the larvæ, and soon a large number of ants were engaged in carrying off the larvæ. When this had continued for about three hours, I raised D $\frac{3}{10}$ of an inch above A. The ants kept on coming and tried hard to reach down from D to A, which was only just out of their reach. Two or three, in leaning over, lost their foothold and dropped into the larvæ; but this was obviously an accident; and after a while they all gave up their efforts, and went away, losing their prize, in spite of most earnest efforts, rather than drop $\frac{3}{10}$ of an inch.

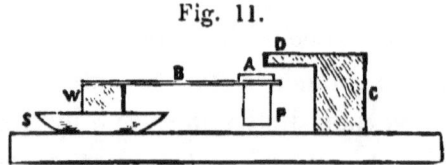

Fig. 11.

At the moment when the separation was made there were fifteen ants on the larvæ. These could, of course, have returned if one had stood still and allowed the others to get on its back. This, however, did not occur to them; nor did they think of letting themselves drop from the bottom of the paper on to the nest. Two or three, indeed, fell down, I have no doubt, by accident; but the remainder wandered about, until at length most of them got into the water. After a time the others abandoned altogether as hopeless the attempt to get at the larvæ.

I waited about six hours, and then again placed the

glass (A) containing the larvæ so as to touch the piece of wood (D), and again put some ants to the larvæ. Soon a regular string of ants was established; when I again raised the wood (D) $\frac{3}{10}$ of an inch above the glass (A), exactly the same result occurred. The ants bent over and made every effort to reach the larvæ, but did not drop themselves down, and after a while again abandoned all hope of getting the larvæ.

In order to test their intelligence, it has always seemed to me that there was no better way than to ascertain some object which they would clearly desire, and then to interpose some obstacle which a little ingenuity would enable them to overcome. Following up, then, the preceding observations, I placed some larvæ in a cup which I put on a slip of glass surrounded by water, but accessible to the ants by one pathway in which was a bridge consisting of a strip of paper $\frac{2}{3}$ inch long and $\frac{1}{3}$ inch wide. Having then put a *Lasius niger* from one of my nests to these larvæ, she began carrying them off, and by degrees a number of friends came to help her. I then, when about twenty-five ants were so engaged, moved the little paper bridge slightly, so as to leave a chasm, just so wide that the ants could not reach across. They came and tried hard to do so; but it did not occur to them to push the paper bridge, though the distance was only about $\frac{1}{3}$ inch, and they might easily have done so. After trying for about a quarter of an hour, they gave up the attempt and returned home. This I repeated several times

Then, thinking that paper was a substance to which they were not accustomed, I tried the same with a bit of straw 1 inch long and $\frac{1}{8}$ inch wide. The result was the same. I repeated this more than once.

Again I suspended some honey over a nest of *Lasius flavus* at a height of about $\frac{1}{8}$ an inch, and accessible only by a paper bridge more than 10 feet long. Under the glass I then placed a small heap of earth. The ants soon swarmed over the earth on to the glass, and began feeding on the honey. I then removed a little of the earth, so that there was an interval of about $\frac{1}{3}$ of an inch between the glass and the earth; but, though the distance was so small, they would not jump down, but preferred to go round by the long bridge. They tried in vain to stretch up from the earth to the glass, which, however, was just out of their reach, though they could touch it with their antennæ; but it did not occur to them to heap the earth up a little, though if they had moved only half a dozen particles of earth they would have secured for themselves direct access to the food. This, however, never occurred to them. At length they gave up all attempts to reach up to the glass, and went round by the paper bridge. I left the arrangement for several weeks, but they continued to go round by the long paper bridge.

Again I varied the experiment as follows:—Having left a nest without food for a short time, I placed some honey on a small wooden brick surrounded by a little moat of glycerine $\frac{1}{8}$ an inch wide and about $\frac{1}{10}$ of an inch in depth. Over this moat I then placed a paper bridge, one end of which rested on some fine mould. I then put an ant to the honey, and soon a little crowd was collected round it. I then removed the paper bridge; the ants could not cross the glycerine; they came to the edge and walked round and round, but were unable to get across, nor did it occur to them to make a bridge or bank across the glycerine with the mould which I had placed so conveniently for them. I was the more surprised at this on account of the ingenuity with which they avail themselves of earth for constructing their nests. For instance, wishing, if possible, to avoid the trouble of frequently moistening the earth in my nests, I supplied one of my communities of *Lasius flavus* with a frame containing, instead of earth, a piece of linen, one portion of which projected beyond the frame and was immersed in water. The linen then sucked up the water by capillary attraction, and thus the air in the frame was kept moist. The ants approved of this arrangement, and took up their quarters in the frame. To minimize evaporation I usually closed the frames all round, leaving only one or two small openings for the ants, but in this case I left the outer side of the frame open. The ants, however, did not like being thus exposed; they therefore brought earth from some little distance, and built up a regular wall along the open side, blocking up the space between the upper and lower plates of glass, and leaving only one or two small openings for themselves. This struck me as very ingenious. The same expedient was, moreover, repeated under similar circumstances by the slave belonging to my nest of *Polyergus*.

The facility or difficulty with which ants find their way, while it partly falls within the section of the subject dealing with their organs of sense, is also closely connected with the question of their general intelligence.

Partly, then, in order to test how far they are guided by sight, partly to test their intelligence, I made various observations and experiments, the accompanying woodcuts being reduced copies of tracings of some of the routes followed by the ants during the course of the observations.

I may here note that the diagrams Figs. 12–17 are careful reductions of large tracings made during the experiments. Though not absolutely correct in every minute detail of contour, they are exact for all practical purposes. As the ants pursued their way, pencil-markings in certain instances, and coloured lines in others, were made so as to follow consecutively the paths pursued.

Experiment 1.—February. On a table communicating with one of my nests (see Fig. 12) I placed upright a common cylindrical lead pencil ¼ inch in diameter and 7 inches long, fastened with sealing-wax to a penny piece. Close to the base of the pencil (A) I brought the end of a paper bridge (B) leading to the nest, and then placed a shallow glass with larvæ at C,

4 inches from the base of the pencil. I then put an ant to the larvæ; when she had become acquainted with the road, she went very straight, as is shown in the woodcut (Fig. 12). In one case, at the point E, she dropped her larva and returned for another. When

Fig. 12.

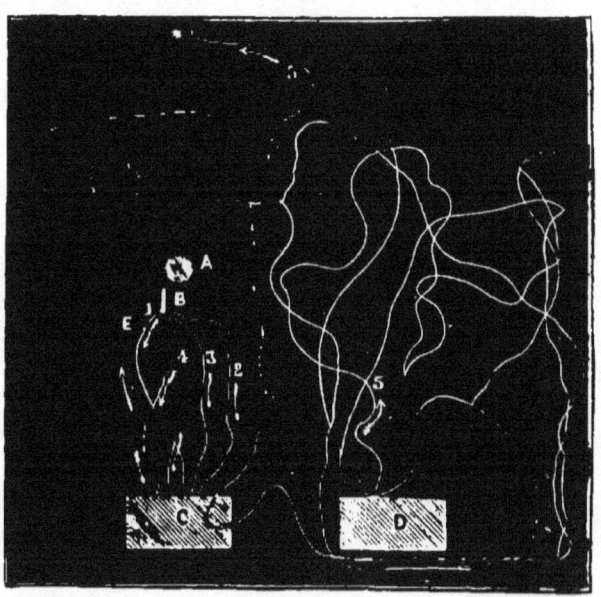

Routes followed in experiment No. 1, as detailed above.

A, position of pencil. B, paper bridge. C and D, glass with larvæ. E, point where larva dropped, the opposite arrow and loop marking return route. 1, 2, 3, 4, comparatively straight paths to the glass. 5, 5, circuitous route on shifting of glass. different access to nest.

she returned on the next journey and was on the glass, I moved it 3 inches, to D, so that the end of the glass was 6 inches from the base of the pencil. If she were much guided by sight, then she would have had little or no difficulty in finding her way back. Her pathway, however (No. 5), which is traced on the paper, shows that she was completely abroad; and, after all, she got back to the nest by a different route.

Fig. 13.

Routes followed in experiment No. 2, as mentioned in text.

B, paper bridge leading to nest. C, glass tray with larvæ, in its first position; and D in its position when shifted. 1, 2, 3, 4, thin white lines indicating the comparatively straight routes. 5, thick white line, and 6, dotted line showing tortuous paths when glass had been altered in position. The arrows indicate directions travelled.

FINDING THEIR WAY. 253

I then varied the experiment as subjoined, and as shown in the woodcut (Fig. 13).

Experiment 2.—I connected the table with the nest by a paper bridge, the end of which is shown at B (Fig. 13), and which came down about an inch from the pole supporting the nest (see Fig. 1). This pole rose 18 inches above the table. I then put the glass tray (C) with larvæ as before, 12 inches from the base of the pole, and put an ant to the larvæ. When she had learnt her way I traced four of her routes, as shown in the thin lines 1, 2, 3, 4. I then on her next journey (5, thick white line), when she was on the tray (C), moved it three inches to D, as shown in the figure, and again traced her route. The contrast is very striking between the relatively straight thin white lines 1, 2, 3, 4 of the four journeys when familiar with the road; whereas in the broad white line No. 5 the zigzag twistings show how much difficulty the ant experienced in finding her way. When she returned I again moved the tray as before, and the dotted sinuous white line (6) shows the course she followed.

Experiment 3.—I then again varied the experiment as follows:—I placed the larvæ in a small china cup on the top of the pencil, which thus formed a column $7\frac{1}{2}$ inches high. The cross line close to the arrows (Fig. 14) is as before, the base of the paper bridge leading to the nest. C shows the position of the penny on which the pencil was supported. The dotted white lines 1, 2, 3, 4 show the routes of a marked ant on four successive journeys from the nest to the base of the pencil. I then moved the pencil 6 inches to D, and the two following routes are marked 5 and 6. In one of them, 5 (thick white line), the ant found a stray

Fig. 14

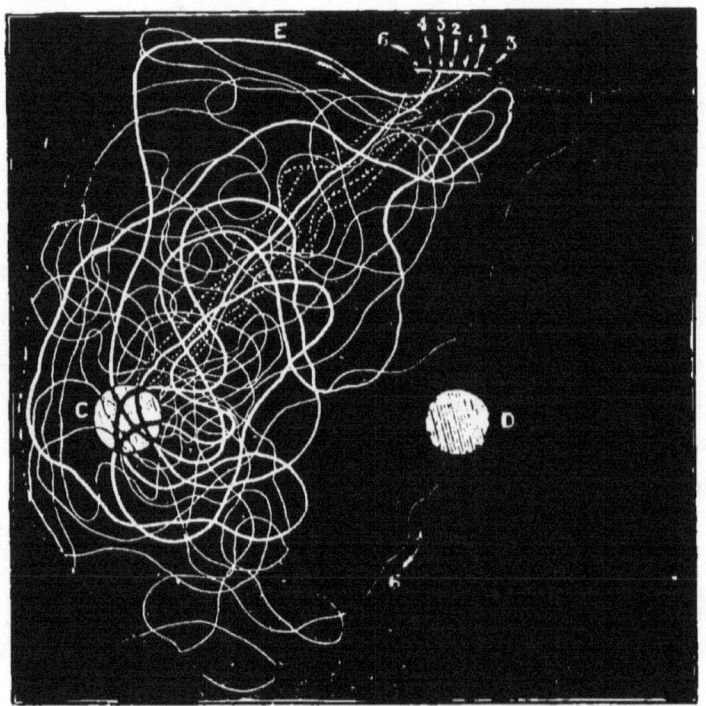

Routes followed in experiment No. 3, as described in text

The line at the six arrows represents a paper bridge going to nest. C, china cup on top of pencil. D, pencil moved. E, place where a stray larva was found. 1, 2, 3, 4, dotted lines show the nearly direct journeys. 5, thick white line (crossing C in black) of route returning to nest, the ant having picked up a stray larva at E. 6, very circuitous thin white line of track from nest to pencil D.

larva at E, with which she returned to the nest, without finding the pencil at all. On the following journey, shown in the fine white zigzag line (6), she found the pencil at last, but, as will be seen, only after many meanderings.

Fig. 15

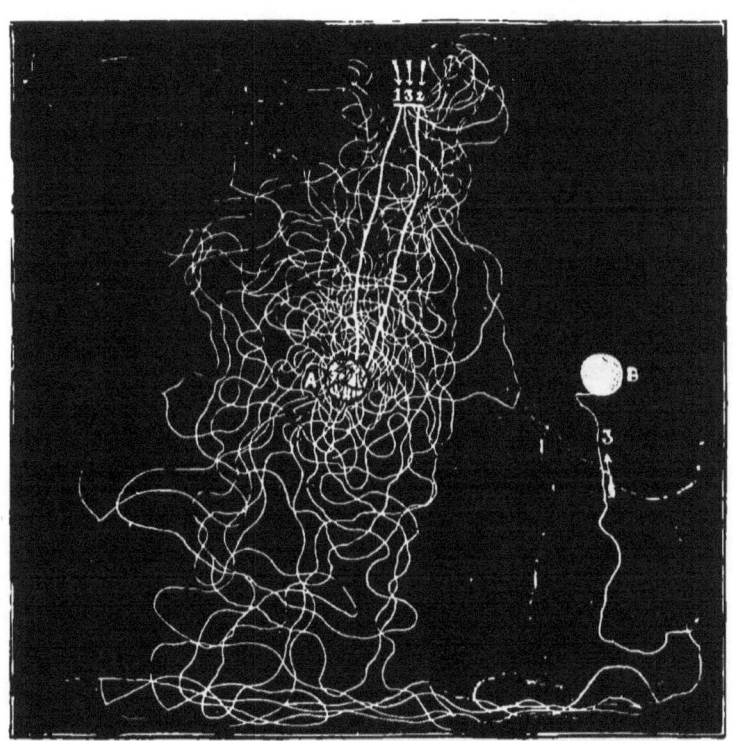

Diagram of complex path traversed in experiment 4.

A, first position of pencil. B, second position of pencil. 1, 2, straight lines of two tracks of the observed ants. 3, winding narrow white line, showing course pursued by the same ant before arriving at B, when the position of the pencil was unchanged.

Experiment 4.—I then repeated the observation

on three other ants (see Figs. 15–17) with the same result: the second was 7 minutes before she found the pencil, and at last seemed to do so accidentally; the third actually wandered about for no less than half an hour (Fig. 15), returning up the paper bridge several times.

Other experiments somewhat similar to the pre-

Fig. 16.

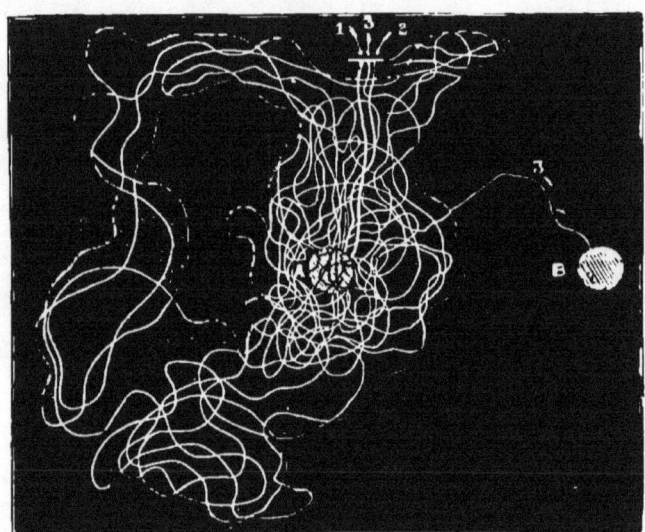

Diagram representing three tracks of an ant in another experiment

A, the first position of pencil and the food, towards which and from the base-line of nest 1 and 2 lead by nearly direct brondish white lines to A. When the latter was removed to B the ant, in its effort to reach this, pursued the narrow white winding line ending in 3 →

ceding, the results of which are shown in the figures 16 and 17, seem to prove that this species of ant, at any

IN FINDING THEIR WAY.   257

rate, guides itself but little by sight. This, which I
had not at all anticipated, seems to follow from the
fact that after the pencil and tray of larvæ had been
removed but a short distance to the right or left, the

Fig. 17

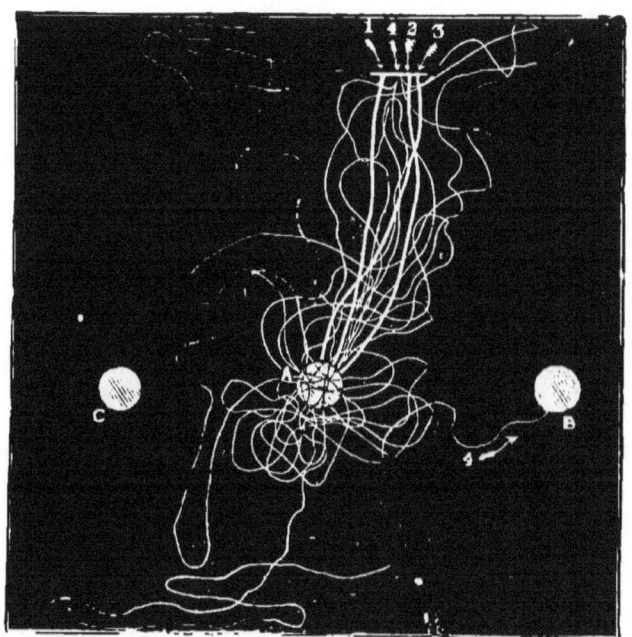

Another tracing showing a similar experiment. 1, 2, 3, the direct
broad lines towards A; and 4, the complicated track made when
reservoir of larvæ was removed to B.

ants on their journey to the shifted object travelled
very often backwards and forwards and around the spot
where the coveted object first stood. Then they would
retrace their steps towards the nest, wander hither and

thither from side to side between the nest and the point A, and only after very repeated efforts around the original site of the larvæ reach, as it were accidentally, the object desired at B.

Another evidence of this consists in the fact that if when ants (*L. niger*) were carrying off larvæ placed in a cup on a piece of board, I turned the board round so that the side which had been turned towards the nest was away from it, and *vice versâ*, the ants always returned over the same track on the board, and, in consequence, directly away from home.

If I moved the board to the other side of my artificial nest, the result was the same. Evidently they followed the road, not the direction.

In order further to test how far ants are guided by sight and how much by scent, I tried the following experiment with *Lasius niger*. Some food was put out at the point a on a board measuring 20 inches by 12 (Fig. 18), and so arranged that the ants in going straight to it from the nest would reach the board at the point b, and after passing under a paper tunnel, c, would proceed between five pairs of wooden bricks, each 3 inches in length and 1¾ in height. When they got to know their way, they went quite straight along the line $d\ e$ to a. The board was

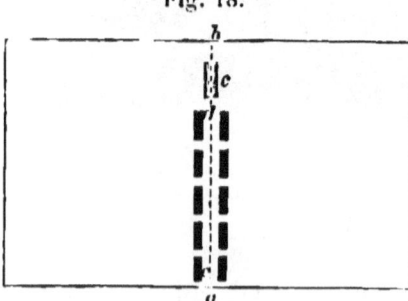

Fig. 18.

then twisted as shown in Fig. 19. The bricks and tunnel being also rearranged so that they were exactly in the same direction as before, but the board having been moved, the line *d e* was now outside them. This change, however, did not at all discompose the ants; but instead of going, as before, through the tunnel and between the rows of bricks to *a*, they walked exactly along the old path to *e*.

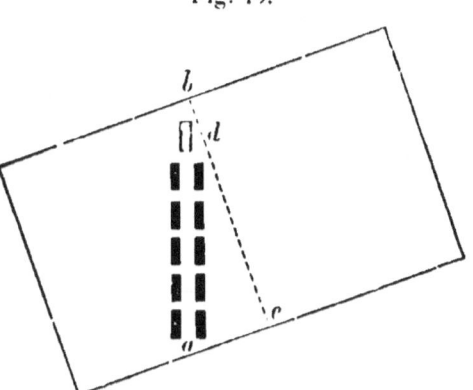

Fig. 19.

I then arranged matters as before, but without the tunnel and with only three pairs of bricks (Fig. 20). When an ant had got quite used to the path *d* to *e*, I altered the position of the bricks and food, as shown in Fig. 21, making a difference of 8 inches in the position of the latter. The ant came as before, walked up to the first brick,

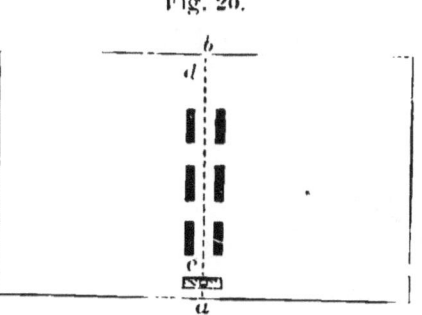

Fig. 20.

touched it with her antennæ, but then followed her old line to *a*. From there she veered towards the food, and very soon found it. When she was gone, I altered it again, as shown in Fig. 22; she returned after the usual interval, and went again straight to a; then, after some wanderings, to f, and at length, but only after a lapse of 25 minutes, found the food at g. These experiments were repeated more than once, and always with similar results. I then varied matters by removing the bricks, which, however, did not seem to make any difference to the ants.

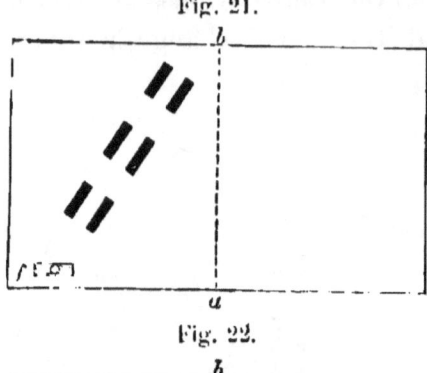

Fig. 21.

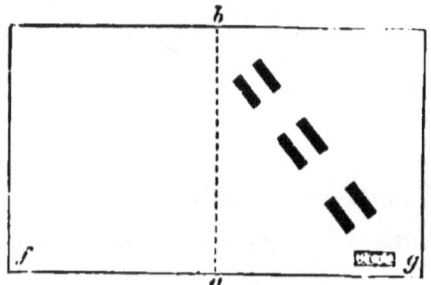

Fig. 22.

I then accustomed some ants (*Lasius niger*) to go to and fro over a wooden bridge, b, c (Fig. 23), to some food.

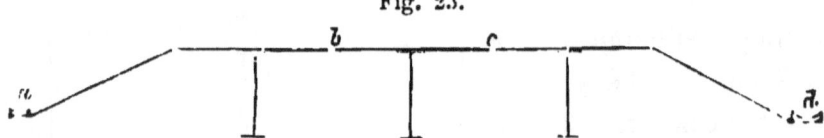

Fig. 23.

When they had got quite accustomed to the way, I watched when an ant was on the bridge and then turned it round, so that the end b was at c, and c at b. In most cases the ant immediately turned round also; but even if she went on to b or c, as the case may be, as soon as she came to the end of the bridge she turned round.

I then modified the arrangement, placing between the nest and the food three similar pieces of wood. Then when the ant was on the middle piece, I transposed the other two. To my surprise this did not at all disconcert them.

I then tried the arrangement shown in Fig. 24.

Fig. 24.

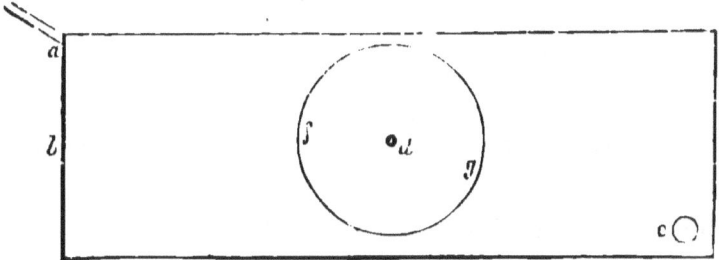

a is a paper bridge leading to the nest; b is a board about 22 inches long by 13 broad, on which is a disk of white paper fastened at the centre by a pin d; e is some food. When the ants had come to know their way so that they passed straight over the paper disk on their way from a to e, I moved the disk round with an ant on it, so that f came to g and g to f. As before, the ants turned round with the paper.

As it might be possible that the ants turned round on account of the changed relative position of external objects, I next substituted a circular box 12 inches in diameter, open at the top, and

7 inches high (in fact, a hat-box) for the flat paper, cutting two small holes at *f* and *g*, so that the ants passing from the nest to the food went through the box entering at *f* and coming out at *g*. The box was fixed

Fig. 25.

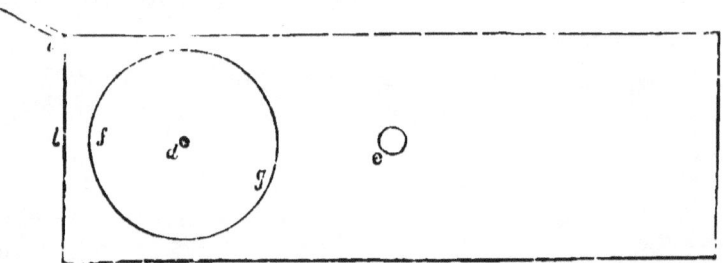

at *d*, so that it might turn easily. I then, when they had got to know their way, turned the box round as soon as an ant had entered it, but in every case the

Fig 26.

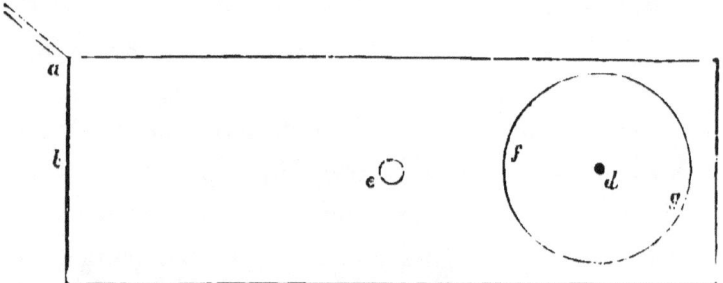

ant turned round too, thus retaining her direction. I then varied the experiment as shown in Figs. 25 and 26.

I replaced the white disk of paper, but put the food *e* at the middle of the board. When the ant had got used to this arrangement I waited till one was on the disk (Fig. 25) and then gently drew it to the other side of *e*, as shown in Fig. 26. In this case, however, the ant did not turn round, but went on to *g*, when she seemed a good deal surprised at finding where she was.

In continuation of the preceding experiments I constructed a circular table 18 inches in diameter. It consisted, as shown in Figs. 27 and 28, of three concentric pieces—a central F G, an intermediate D E, H I, and an outer piece B C, K L, each of these

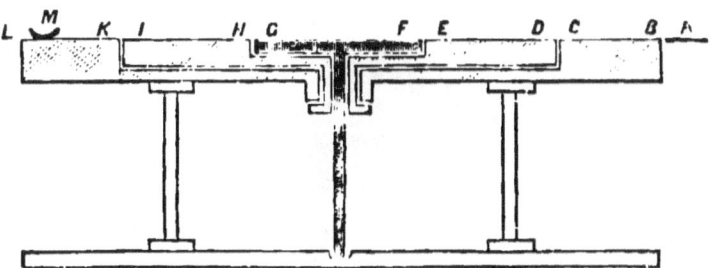

Fig. 27.

three pieces being capable of separate rotation. This arrangement was kindly devised for me by Mr. Francis Galton.

I then connected the table with a nest of *Lasius niger* by a paper bridge A, and also made a paper path across the table, as shown in Fig. 28, divided into five pieces corresponding to the divisions of the table. This I did because I found that the ants wandered less if they were provided with a paper road than if they walked actually on the wood itself. I then placed a cup containing larvæ on the table at B, and put an ant on

264   EXPERIMENTS WITH

the larvæ. She at once picked one up, and, with some little guidance from me, carried it off to the nest, returning at once for another, bringing some friends with her to help. When she knew her way, I gradually moved the cup across the table along the paper path

Fig. 28.

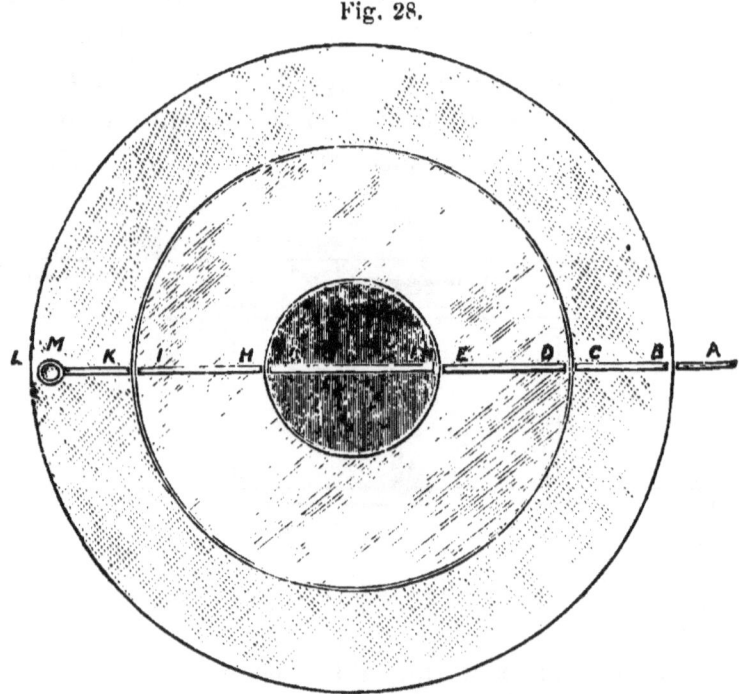

to M, placing it on a column five inches high. After a while the ants came to know the way quite well, and passed straight along the path from the nest to the larvæ at M. Having thus established a service of ants, I tried the following experiments:—

1. I removed the piece of paper G I. This dis-

turbed them; but they very soon re-established the chain.

2. I turned round the central piece of the table G F, so that the paper G F was reversed, G being where F had been, and *vice versâ*. This did not seem to diconcert the ants at all. They went straight over the paper as before, without a moment's hesitation.

3. When some ants were between I and D, I rotated the outer circle of the table halfway round, which of course carried the cup containing the larvæ from L to B. The ants took no notice of this, but went straight to L.

4. When some ants were between I and D, I rotated the table several times, bringing it finally to its original position. This disturbed them a good deal; but eventually they all continued their course to L.

5. When some ants were between I and D, I half rotated the two centre parts of the table, the result of which, of course, was that the ant was moving towards, instead of away from, the nest. In every case the ants turned round too, so as duly to reach L. So also those which were on their way back from the larvæ to the nest turned in the same manner.

6. When the ants were between I and D, I half rotated the whole table. Again the ants turned round too, though of course in this case, when they reached the place where L had been, the cup with the larvæ was behind them at B.

The two latter experiments, though quite in accordance with those previously made, puzzled me a good deal. Experiment 3, as well as some of those recorded previously, seemed to show that ants were little guided in such cases by the position of surrounding objects. However, I was anxious to test this.

7. Accordingly I took a round box and placed it upside down on the table, having cut two niches, one at each side, where it lay on the paper path, so as to afford a passage for the ants, as in the experiments recorded in my previous paper; but on this occasion I left the lid on, cutting, however, a hole through which I could watch the result. In this case, therefore, the surrounding objects, *i.e.* the walls of the box, turned round with the table. Then, as before, when the ants were between I and D, I turned the table half round. The results were as follows:—

	Ants which turned	Ants which did not turn
Experiment 1	1	2
,, 2	1	1
,, 3	1	1
,, 4	4	2
,, 5	0	1
,, 6	0	1
,, 7	0	3
,, 8	1	1
,, 9	0	1
,, 10	2	2
,, 11	1	1
,, 12	0	3
	11	19

In this case, then, only 11 ants turned; and as 4 of them were together, it is possible that 3 simply followed the first. Moreover, the ants which turned did so with much more hesitation and less immediately.

8. For comparison, I then again tried the same experiment, but without the box. The results were as follows:—

	Ants which turned	Ants which did not turn
Observation 1	3	0
,, 2	3	0
,, 3	3	1 ?
,, 4	3	0
,, 5	4	0
,, 6	4	0
	20	1

Under these circumstances, therefore, all the ants but one certainly turned, and her movements were undecided.

From these last two experiments it is obvious that the presence of the box greatly affected the result, and yet the previous results made it difficult to suppose that the ants noticed any objects so distant as the walls of the rooms, or even as I was myself. The result surprised me considerably; but I think the explanation is given by the following experiments.

I again put some larvæ in a cup, which I placed in the centre of the table; and I let out an ant which I had imprisoned after the previous experiments, placing her in the cup; she carried off a larva to the nest and soon returned. When she was again in the cup I half rotated the table: when she came out she seemed a little surprised; but after walking once round the cup, started off along the paper bridge straight home. When she returned to the cup I again half rotated the table. This time she went back quite straight When she had come again, I once more half rotated the table; she returned quite straight. Again the same happened. A second ant then came: I half rotated the table as before. She went wrong for about an inch and a half, but then turned round and went straight home.

I was working by the light of two candles which were on the side of the table towards the nest. The next time the two ants came I half rotated the table as before, and moved the candles to the far side. This time the ants were deceived, and followed the paper bridge to the end of the table furthest from the nest. This I repeated a second time, with the same result. I then turned the table as before without altering the lights, and the ants (four of them) went back all right. I then again turned the table, altering the lights, and the ant went wrong.

I then altered the lights without rotating the table. the first ant went wrong; the second right; the third wrong; the fourth wrong; the fifth hesitated some seconds, and then went wrong; the sixth right; the seventh went all but to the edge the wrong way, but, after various wanderings, at last went right. When, therefore, the direction of the light was changed, but everything else left as before, out of seven ants, five were deceived and went in the wrong direction.

After an interval of a week, on March 25, I arranged

the nest and the rotating table as before, and let out three ants which I had imprisoned on the 19th, and which knew their way. I put them on the larvæ at M as before. The paper pathway had been left untouched. The ants examined the larvæ and then went straight home along the paper path; but, to my surprise, only one of them carried off a larva. Nevertheless they had evidently taken the news to the nest, for the ants at once began coming to the cup in considerable numbers and carrying off the larvæ. I do not altogether understand this proceeding, and unluckily had not marked the first three ants; so that I cannot tell whether they brought or sent their friends. It seems possible that they felt unequal to the exertion of carrying a burthen to the nest until they had had some food.

When the ants were fairly at work I turned the table 90 degrees. In this case eight ants which were on their way to the larvæ continued their march along the paper, while two turned back; but none left the paper and went across the table straight for the larvæ.

I then stopped the experiment for a while, so that the excitement might subside; as when the ants become too numerous it is not so easy to watch them.

When all was quiet I put the cup with the larvæ on the middle of the table, and covered the greater part of the table with the box as before. In a short time some ants again came to the larvæ, and then, just as they were leaving the cup on their way home, I turned the table, as before, half round.

Under these circumstances, however, instead of

turning as in the previous experiment, ten ants, one after another, continued their course, thus coming out of the box at the end furthest from the nest. When ten ants successively had, under these circumstances, gone wrong, to make the experiment complete, I tried it again, everything being the same, except that there was no box. Under these circumstances five ants, one after the other, turned directly the table was rotated.

From these experiments, therefore, it seems clear that in determining their course the ants are greatly influenced by the direction of the light.

March 27.—I let out two ants imprisoned on the 25th, and placed them on the larvæ, which I put on a column 7 inches high, covered with blue paper, and communicating with the nest by the paper path (A, Fig. 29) arranged as usual, but supported on pins. At first I arranged it as shown below, placing the larvæ at M, on a table 18 inches in diameter, so that the ants, on arriving at the larvæ, made nearly a semi-circle round the edge of the table. I then gradually moved the larvæ to M' and afterwards to M". The ants, however, obviously knew that they were going unnecessarily round. They ran along the paper bridge in a very undecided manner, continually turning round and often coming down the

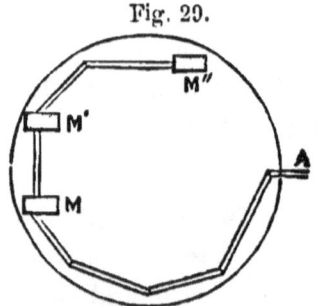

Fig. 29.

pins; while in returning to the nest they persistently came down the side of the pillar nearest to the nest, though I repeatedly attempted to guide them the other way. Even when placed on the paper bridge between M and M', they were very dissatisfied. In fact, it was obvious that they knew they were being sent a long way round, and were attempting to make a shorter cut.

I then again placed the larvæ on the column at M, and when the ants were once more going to and fro regularly along the paper path, I altered the position of the column and larvæ to M', placing the edge of the pillar, which the ants had been accustomed to ascend, towards the paper bridge, connecting it with the original bridge by a side-bridge a, M being an inch from the original bridge. Under these circumstances three ants ran on to M; then two found their way over the bridge a to M'. Of the next ten ants, five went to M and five over a to M'. The next ten all went over the paper bridge a to M'.

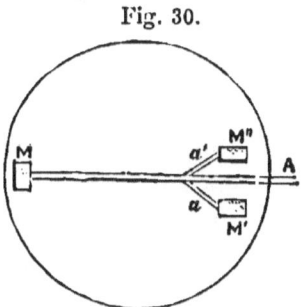

Fig. 30.

I then put the pillar and the larvæ on the other side of the original paper path at M'', connected with the main path by a short bridge a', taking for a' a new piece of paper, so that scent would be no guide. I left the little bridge a in its place. The ants went as follows:—

To M″	1	To M′	0	To M	0
,,	1	,,	0	,,	1
,,	1	,,	0	,,	1
,,	1	,,	0	,,	1
,,	1	,,	1	,,	1
,,	0	,,	0	,,	1
,,	1	,,	0	,,	0
,,	1	,,	0	,,	0
,,	1	,,	0	,,	0
,,	1	,,	0	,,	0
,,	1	,,	1	,,	0
,,	1	,,	1	,,	0
,,	1	,,	0	,,	0
	12		3		5

It seems clear, therefore, that though the ants did not trust so much to their eyes as a man would have done under similar circumstances, yet that they were to some extent guided by sight.

I then removed all the paper pathways and put the pillar to M. Of the first two ants which came to the table, the first found the pillar in five minutes, the second, after wandering about for a quarter of an hour, gave the search up in despair, and went home. I then moved the pillar to M′, and watched the next ant that came on to the table; she found it in a minute or

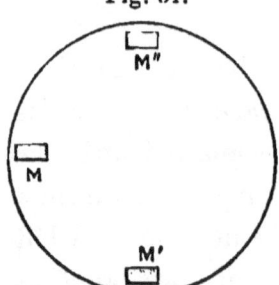

Fig. 31.

two. I then moved it to M". Two ants came together. One found the pillar in 7 minutes; the other took no less than 25, although, as already mentioned, the table was only 18 inches in diameter. Obviously, therefore, though it seems clear that they are helped by sight, still these last observations support those previously recorded, and show that in finding their way they do not derive by any means so much assistance from their eyes as we should under corresponding circumstances

CHAPTER X.

BEES.

I ORIGINALLY intended to make my experiment principally with bees, but soon found that ants were on the whole more suitable for my purpose.

In the first place, ants are much less excitable, they are less liable to accidents, and from the absence of wings are more easy to keep under continuous observation.

Still, I have made a certain number of observations with bees, some of which may be worth here recording.

As already mentioned, the current statements with reference to the language of social insects depend much on the fact that when one of them, either by accident or in the course of its rambles, has discovered a stock of food, in a very short time many others arrive to profit by the discovery. This, however, does not necessarily imply any power of describing localities. If the bees or ants merely follow their more fortunate comrade, the matter is comparatively simple; if, on the contrary, others are sent, the case becomes very different.

In order to test this I proposed to keep honey in a given place for some time, in order to satisfy myself that it would not readily be found by the bees; and then, after bringing a bee to the honey, to watch whether it brought others, or sent them—the latter of course implying a much higher order of intelligence and power of communication.

I therefore placed some honey in a glass, close to an open window in my sitting-room, and watched it for sixty hours of sunshine, during which no bees came to it.

I then, at 10 A.M. on a beautiful morning in June, went to my hives, and took a bee which was just starting out, brought it in my hand up to my room (a distance of somewhat less than 200 yards), and gave it some honey, which it sucked with evident enjoyment. After a few minutes it flew quietly away, but did not return; nor did any other bee make its appearance.

The following morning I repeated the same experiment. At 7.15 I brought up a bee, which sipped the honey with readiness, and after doing so for about four minutes flew away with no appearance of alarm or annoyance. It did not, however, return; nor did any other bee come to my honey.

On several other occasions I repeated the same experiments with a like result. Altogether I tried it more than twenty times. Indeed, I rarely found bees to return to honey if brought any considerable distance at once. By taking them, however, some twenty yards each time they came to the honey, I at length trained them to come to my room. On the whole, however, I found it more convenient to procure one of Marriott's observatory hives, both on account of its construction, and also because I could have it in my room, and thus keep the bees more immediately under my own eye. My room is square, with three windows, two on the south-west side, where the hive was placed, and one on the south-east. Besides the ordinary entrance from the outside, the hive had a small postern door opening into the room; this door was provided with an alighting-board, and closed by a plug: as a general rule the bees did not notice it much unless the passage was very full of them.

I then placed some honey on a table close to the hive, and from time to time fed certain bees on it. Those which had been fed soon got accustomed to come for the honey; but partly on account of my frequent absence from home, and partly from their difficulty in finding their way about, and their tendency to lose themselves, I could never keep any marked bee under observation for more than a few days.

Out of a number of similar observations I will here mention a few and give them in detail in the Appendix, as throwing some light on the power of communicating facts possessed by the bees; they will also illustrate the daily occupations of a working bee.

Experiment 1.—Thus, on August 24, 1874, I opened the postern door leading into my room at 6.45 A.M., and watched till 1 P.M. three bees, which had been trained to come to honey at a particular spot. They did not, however, know their way very well, and consequently lost a good deal of time. One made 23 journeys backwards and forwards between the hive and the honey, the second 13, and the third only 7.

The following day I watched the first of these bees from 7.23 to 12.54, during which time she made 19 journeys. Scarcely any other bees came, but I did not record the exact number.

Experiment 2.—I watched another bee from 6.55 A.M. till 7.15 P.M., during which time she made 59 visits to the honey, and only one other bee came to it.

Experiment 3.— Another from 7 A.M. till 3 P.M.; she made 40 journeys, and only two other bees came. She returned the two following mornings, and was watched for three hours each day, during which time no other bee came.

Experiment 4.—Another morning I watched a different bee from 9.19 A.M. to 2 P.M.: she made 21 journeys, and no other bee came.

Then, thinking that perhaps this result might be due to the quantity of honey being too small, I used a wide-mouthed jar, containing more than one pound of honey.

Experiment 5.—I watched two bees from 1.44 till 4.30, during which time they made 24 journeys, but only one other bee came.

Experiment 6.— Besides the honey in the jar I spread some out over two plates, so as to increase the surface. I watched a bee from 12.15 till 6.15 P.M. She made 28 journeys, but did not bring a single **friend** with her.

Experiment 7.—On July 19 I put a bee to a honeycomb which contained twelve and a half pounds of honey at 12.30, and which was placed in a corner of my room as far as possible from the window. That afternoon she made 22 visits to it, and no other bee came. The following morning she returned at 6.5 A.M., and I watched her till 2. She made 22 journeys, but did not bring a single friend with her.

Experiment 8.—Another bee was also brought to the same honeycomb, watched from 2.30 till 7.14. She made 14 journeys, but did not bring a single friend.

I might give other similar cases, but these are, I think, sufficient to show that bees do not bring their friends to share any treasure they have discovered, so invariably as might be assumed from the statements of previous observers. Possibly the result is partly due to the fact that my room is on the first floor, so that the bees coming to it flew at a higher level than that generally used by their companions, and hence were less likely to be followed.

Indeed, I have been a good deal surprised at the difficulty which bees experience in finding their way.

For instance, I put a bee into a bell-glass 18 inches long, and with a mouth $6\frac{1}{2}$ inches wide, turning the closed end to the window; she buzzed about for an hour, when, as there seemed no chance of her getting out, I put her back into the hive. Two flies, on the contrary, which I put in with her, got out at once. At 11.30 I put another bee and a fly into the same glass: the latter flew out at once. For half an hour the bee tried to get out at the closed end; I then turned the glass with its open end to the light, when she flew out at once. To make sure, I repeated the experiment once more with the same result.

Some bees, however, have seemed to me more intelligent in this respect than others. A bee which I had fed several times, and which had flown about in the room, found its way out of the glass in a quarter of an hour, and when put in a second time came out at once. Another bee, when I closed the postern door which opened from my hive directly into my room, used to come round to the honey through an open window.

One day (April 14, 1872), when a number of them were very busy on some berberries, I put a saucer with some honey between two bunches of flowers; these flowers were repeatedly visited, and were so close that there was hardly room for the saucer between them, yet from 9.30 to 3.30 not a single bee took any notice of the honey. At 3.30 I put some honey on one of the bunches of flowers, and it was eagerly sucked by the bees; two kept continually returning till past five in the evening.

One day when I came home in the afternoon I found that at least a hundred bees had got into my room through the postern and were on the window, yet not one was attracted by an open jar of honey which stood in a shady corner about 3 feet 6 inches from the window.

Another day (April 29, 1872) I placed a saucer of honey close to some forget-me-nots, on which bees were numerous and busy; yet from 10 A.M. till 6 only one bee went to the honey.

I put some honey in a hollow in the garden wall opposite my hives at 10.30 (this wall is about five feet high and four feet from the hives), yet the bees did not find it during the whole day.

On March 30, 1873, a fine sunshiny day, when the bees were very active, I placed a glass containing honey at 9 in the morning on the wall in front of the hives; but not a single bee went to the honey the whole day. On April 20 I tried the same experiment with the same result.

September 19.—At 9.30 I placed some honey in a glass about four feet from and just in front of the hive, but during the whole day not a bee observed it.

As it then occurred to me that it might be suggested that there was something about this honey which rendered it unattractive to the bees, on the following day I first placed it again on the top of the wall for three hours, during which not a single bee came, and then moved it close to the alighting-board of the hive. It remained unnoticed for a quarter of an hour, when two bees observed it, and others soon followed in considerable numbers.

It is generally stated not only that the bees in a hive all know one another, but also that they immediately recognise and attack any intruder from another hive. It is possible that the bees of particular hives have a particular smell. Thus Langstroth, in his interesting 'Treatise on the Honey-Bee,' says, 'Members of different colonies appear to recognise their hive companions by the sense of smell;' and I believe that if colonies are sprinkled with scented syrup they may generally be safely mixed. Moreover, a bee returning to its own hive with a load of treasure is a very different creature from a hungry marauder; and it is said that a bee, if laden with honey, is allowed to enter any hive with impunity. Mr. Langstroth continues: 'There is an air of roguery about a thieving bee which, to the expert, is as characteristic as are the motions of a pickpocket to a skilful policeman. Its sneaking look and nervous, guilty agitation, once seen, can never be mistaken.' It is at any rate natural that a bee which enters a wrong hive by accident should be much surprised and alarmed, and would thus probably betray herself.

So far as my own observations go, though bees habitually know and return to their own hive, still, if placed on the alighting-board of another, they often enter it without molestation. Thus:—

On May 4 I put a strange bee into a hive at 2 o'clock. She remained in till 2.20, when she came out, but entered again directly. I was away most of the afternoon, but returned at 5.30; at 6 she came out of the hive, but soon returned; and after that I saw no more of her.

May 12.—A beautiful day, and the bees very active. I placed twelve marked bees on the alighting-board of a neighbouring hive. They all went in; but before evening ten had returned home.

May 13.—Again put twelve marked bees on the alighting-board of another nest; eleven went in. The following day I found that seven had returned home; the other five I could not see.

May 17.—Took a bee, and, after feeding her and marking her white, put her to a hive next but one to her own at 4.18. She went in.

4.22. Came out and went in again.
4.29. Came out. I fed her and sent her back.
4.35. Came out. Took a little flight and came back.
4.45. Went in, but returned.   4.52. Went in.
4.53. Came out.   4.56. ,,
4.57. ,,   4.58. ,,
5. 1. Came out, took another little flight, and returned. I fed her again.   5.25. Went in again.
5.28. Came out again.   5.29. ,,
5.31. ,,   5.33. ,,
5.36. ,,   5.40. ,,
5.46. Shut her and the others in with a piece of note-paper.
6.36. One of the bees forced her way through. I opened the door; and several, including the

white one, came out directly. Till 6.50 this bee kept on going in and out every minute or two; hardly any bees were flying, only a few standing at the doors of most of the hives. At 7.20 she was still at the hive door.

May 20.—Between 6 and 7 P.M. I marked a bee and transferred her to another hive.

May 21.—Watched from 7.30 to 8.9 in the morning without seeing her. At half-past six in the evening went down again, directly saw and fed her. She was then in her new hive; but a few minutes after I observed her on the lighting-stage of her old hive; so I again fed her, and when she left my hand she returned to the new hive.

May 22.—8 o'clock. She was back in her old hive.

May 23.—About 12.30 she was again in the new hive.

Though bees which have stung and lost their sting always perish, they do not die immediately; and in the meantime they show little sign of suffering from the terrible injury. On August 25 a bee which had come several times to my honey was startled, flew to one of the windows, and had evidently lost her way. While I was putting her back she stung me, and lost her sting in doing so. I put her in through the postern, and for twenty minutes she remained on the landing-stage; she then went into the hive, and after an hour returned to the honey and fed quietly, notwithstanding the terrible injury she had received. After this, however, I did not see her any more.

Like many other insects, bees are much affected by light. One evening, having to go down to the cellar, I lit a small covered lamp. A bee which was out came to it, and, flying round and round like a moth, followed me the whole of the way there.

I often found that if bees which were brought to honey did not return at once, still they would do so a day or two afterwards. For instance, on July 11, 1874, a hot thundery day, and when the bees were much out of humour, I brought twelve bees to some honey: only one came back, and that one only once; but on the following day several of them returned.

My bees sometimes ceased work at times when I could not account for their doing so. October 19 was a beautiful, sunshiny, warm day. All the morning the bees were fully active. At 11.25 I brought one to the honeycomb, and she returned at the usual intervals for a couple of hours; but after that she came no more, nor were there any other bees at work. Yet the weather was lovely, and the hive is so placed as to catch the afternoon sun.

I have made a few observations to ascertain, if possible, whether the bees generally go to the same part of the hive. Thus,--

October 5.—I took a bee out of the hive, fed her, and marked her. She went back to the same part.

October 9.—At 7.15 I took out two bees, fed and

marked them. They returned; but I could not see them in the same part of the hive. One, however, I found not far off.

At 9.30 brought out four bees, fed and marked them. One returned to the same part of the hive. I lost sight of the others.

Since their extreme eagerness for honey may be attributed rather to their anxiety for the commonweal than to their desire for personal gratification, it cannot fairly be imputed as greediness; still the following scene, described by Dr. Langstroth, and one which most of us have witnessed, is incompatible surely with much intelligence. 'No one can understand the extent of their infatuation until he has seen a confectioner's shop assailed by myriads of hungry bees. I have seen thousands strained out from the syrup in which they had perished; thousands more alighting even upon the boiling sweets; the floor covered and windows darkened with bees, some crawling, others flying, and others still so completely besmeared as to be able neither to crawl nor fly—not one in ten able to carry home its ill-gotten spoils, and yet the air filled with new hosts of thoughtless comers.'[1]

If, however, bees are to be credited with any moral feelings at all, I fear the experience of all bee-keepers shows that they have no conscientious scruples about robbing their weaker brethren. 'If the bees of a strong stock,' says Langstroth, 'once get a taste of forbidden

[1] *Hive- and Honey-Bee*, Langstroth, p. 277.

sweets, they will seldom stop until they have tested the strength of every hive.' And again, 'Some bee-keepers question whether a bee that once learns to steal ever returns to honest courses.' Siebold has mentioned similar facts in the case of certain wasps (*Polistes*).

Far, indeed, from having been able to discover any evidence of affection among them, they appear to be thoroughly callous and utterly indifferent to one another. As already mentioned, it was necessary for me occasionally to kill a bee; but I never found that the others took the slightest notice. Thus on October 11 I crushed a bee close to one which was feeding—in fact, so close that their wings touched; yet the survivor took no notice whatever of the death of her sister, but went on feeding with every appearance of composure and enjoyment, just as if nothing had happened. When the pressure was removed, she remained by the side of the corpse without the slightest appearance of apprehension, sorrow, or recognition. She evidently did not feel the slightest emotion at her sister's death, nor did she show any alarm lest the same fate should befall her also. In a second case exactly the same occurred. Again, I have several times, while a bee has been feeding, held a second bee by the leg close to her; the prisoner, of course, struggled to escape, and buzzed as loudly as she could; yet the bee which was feeding took no notice whatever. So far, therefore, from being at all affectionate, I doubt whether bees are in the least fond of one another

Their devotion to their queen is generally quoted as an admirable trait; yet it is of the most limited character. For instance, I was anxious to change one of my black queens for a Ligurian; and accordingly on October 26 Mr. Hunter was good enough to bring me a Ligurian queen. We removed the old queen, and we placed her with some workers in a box containing some comb. I was obliged to leave home on the following day; but when I returned on the 30th I found that all the bees had deserted the poor queen, who seemed weak, helpless, and miserable On the 31st the bees were coming to some honey at one of my windows, and I placed this poor queen close to them. In alighting, several of them even touched her; yet not one of her subjects took the slightest notice of her. The same queen, when afterwards placed in the hive, immediately attracted a number of bees.

As regards the affection of bees for one another, it is no doubt true that when they have got any honey on them, they are always licked clean by the rest; but I am satisfied that this is for the sake of the honey rather than of the bee. On September 27, for instance, I tried with two bees: one 'had been drowned, the other was smeared with honey. The latter was soon licked clean; of the former they took no notice whatever. I have, moreover, repeatedly placed dead bees by honey on which live ones were feeding, but the latter never took the slightest notice of the corpses.

Dead bees are indeed usually carried out of the

hive, but if one is placed on the alighting-stage, the others seem to take no notice of it, though it is in general soon pushed off accidentally by their movements. I have even seen the bees sucking the juices of a dead pupa.

As regards the senses of bees, it seems clear that they possess a keen power of smell.

On October 5 I put a few drops of eau de Cologne in the entrance of one of my hives, and immediately a number of bees (about fifteen) came out to see what was the matter. Rose-water also had the same effect; and, as will be mentioned presently, in this manner I called the bees out several times; but after a few days they took hardly any notice of the scent.

These observations were made partly with the view of ascertaining whether the same bees act as sentinels. With this object, on October 5 I called out the bees by placing some eau de Cologne in the entrance, and marked the first three bees that came out. At 5 P.M. I called them out again; about twenty came, including the three marked ones. I marked three more.

October 6.— Called them out again. Out of the first twelve, five were marked ones. I marked three more.

October 7.—Called them out at 7.30 A.M. as before. Out of the first nine, seven were marked ones.

At 5.30 P.M. called them out again. Out of six, five were marked ones.

October 8.—Called them out at 7.15. Six came out, all marked ones.

October 9.—Called them out at 6.40. Out of the first ten, eight were marked ones.

Called them out at 11.30 A.M. Out of six, three were marked. I marked the other three.

Called them out at 1.30 P.M. Out of ten, six were marked.

Called them out at 4.30. Out of ten, seven were marked.

October 10.—Called them out at 6.5 A.M. Out of six, five were marked.

Shortly afterwards I did the same again, when out of eleven, seven were marked ones.

5.30 P.M. Called them out again. Out of seven, five were marked.

October 11.—6.30 A.M. Called them out again. Out of nine, seven were marked.

5 P.M. Called them out again. Out of seven, five were marked.

After this day they took hardly any notice of the scents.

Thus in these nine experiments, out of the ninety-seven bees which came out first, no less than seventy-one were marked ones, though out of the whole number of bees in the hive there were only twelve marked for this purpose, and, indeed, even fewer in the earlier experiments. I ought, perhaps, to add that I generally fed the bees when I called them out.

The Sense of Hearing.

August 29.—The result of my experiments on the hearing of bees has surprised me very much. It is generally considered that to a certain extent the emotions of bees are expressed by the sounds they make,[1] which seems to imply that they possess the power of hearing. I do not by any means intend to deny that this is the case. Nevertheless I never found them take any notice of any noise which I made, even when it was close to them. I tried one of my bees with a violin. I made all the noise I could, but to my surprise she took no notice. I could not even see a twitch of the antennæ. The next day I tried the same with another bee, but could not see the slightest sign that she was conscious of the noise. On August 31 I repeated the same experiment with another bee with the same result. On September 12 and 13 I tried several bees with a dog-whistle and a shrill pipe; but they took no notice whatever, nor did a set of tuning-forks which I tried on a subsequent day have any more effect. These tuning-forks extended over three octaves, beginning with *a* below the ledger line. I also tried with my voice, shouting, &c., close to the head of a bee; but, in spite of my utmost efforts, the bees took no notice. I repeated these experiments at night when the bees were quiet; but no noise that I could make seemed to disturb them in the least.

[1] See, for instance, Landois, *Zeits. f. wiss. Zool.* 1867, p. **184**

In this respect the results of my observations on bees entirely agreed with those on ants, and I will here, therefore, only refer to what has been said in a preceding chapter.

The Colour Sense of Bees.

The consideration of the causes which have led to the structure and colouring of flowers is one of the most fascinating parts of natural history. Most botanists are now agreed that insects, and especially bees, have played a very important part in the development of flowers. While in many plants, almost invariably with inconspicuous blossoms, the pollen is carried from flower to flower by the wind, in the case of almost all large and brightly coloured flowers this is effected by the agency of insects. In such flowers the colours, scent, and honey serve to attract insects, while the size and form are arranged in such a manner that the insects fertilise them with pollen brought from another plant.

There could, therefore, be little doubt that bees possess a sense of colour. Nevertheless I thought it would be desirable to prove this if possible by actual experiment, which had not yet been done. Accordingly on July 12 I brought a bee to some honey which I placed on blue paper, and about 3 feet off I placed a similar quantity of honey on orange paper. After she had returned twice I transposed the papers; but she returned to the honey on the blue paper. After she had made three more visits, always to the blue paper, I transposed them again, and she again followed the colour, though the honey was left in the same place. The following day I was not able to watch her; but on the 14th at—

7.29 A.M. she returned to the honey on the blue paper
7.31 left.
7.44 ,, ,, 7.41 ,,
7.56 ,, ,,

I then again transposed the papers. At 8.5 she returned to the old place, and was just going to alight; but observing the change of colours, without a moment's hesitation darted off to the blue. No one who saw her at that moment could have entertained the slightest doubt about her perceiving the difference between the two colours. At 8.9 she went.

8.13 she returned to the blue; 8.16 went.
8.20 ,, ,, 8.23 ,,
8.26 ,, ,, 8.30 ,,

Transposed the colours again.

At 8.35 she returned to the blue, and at 8.39 went.
8.44 ,, ,, 8.47 ,,
8.50 ,, ,, 8.53 ,,

Transposed the colours again.

8.57 she returned again to the blue; 9. 0 ,,
9. 4 ,, ,, 9. 7 ,,
9.12 ,, ,, 9.15 ,,

COLOURED PAPER.

9.19 she returned again to the blue; 9.22 went.
9.25 ,, ,, 9.27 ,,
9.30 ,, ,, 9.34 ,,
9.40 ,, ,, 9.44 ,,
9.50 ,, ,, 9.55 ,,

Transposed the colours again.

10. 2 she returned again to the blue; 10. 6 ,,
10.10 ,, ,, 10.14 ,,
10.20 ,, ,, 10.25 ,,
10.30 ,, ,, 10.34 ,,
10.40 ,, ,, 10.44 ,,
10.48 ,, ,, 10.51 ,,
11.12 ,, ,, 11.14 ,,
11.21 ,, and flew about, having been disturbed.
11.26 ,, ,, 11.28 went.
11.36 ,, ,, 11.40 ,,
12. 5 came and flew about, but did not settle till—
12.17 she returned again to the blue; 12.17 went.
12.21 came and flew about.

Though it was a beautiful afternoon, she did not return any more that day.

On October 2 I placed some honey on slips of glass resting on black, white, yellow, orange, green, blue, and red paper. A bee which was placed on the orange returned twenty times to that slip of glass, only once or twice visiting the others, though I moved the position and also the honey. The next morning again two or three bees paid twenty-one visits to the orange and yellow, and only four to all the other slips of glass. I then moved the glass, after which, out of thirty-two visits, twenty-two were to the orange and yellow. This was due, I believe, to the bee having been placed on the orange at the beginning of the experiment. I do not attribute it to any preference for the orange or yellow; indeed, I shall presently give reasons for considering that blue is the favourite colour of bees.

October 6.—I had ranged my colours in a line, with the blue at one end. It was a cold morning, and only one bee came. She had been several times the preceding day, generally to the honey which was on the blue paper. This day also she came to the blue; I moved the blue gradually along the line one stage every half-hour, during which time she paid fifteen visits to the honey, in every case going to that which was on the blue paper.

Again, on September 13 at 11 A.M., I brought up a bee from one of my hives; at 11.40 she returned to honey which I had put on a slip of glass on green paper. She returned at 11.51. And again at

	12. 1
,,	12.13
,,	12.22
,,	12.33
,,	12.46
,,	12.58

DISTINGUISHING COLOURS. 295

She returned at 1.12. This time she lost her way in the room.
,, 1.49
,, 2. 1. This time she got stuck in the honey, and had to clean herself.
,, 2.25
,, 2.40. I now put red paper instead of the green, and put the green paper with a similar quantity of honey on it a foot off.
,, 2.51 to the honey on green paper. I then gently moved the green paper, with the bee on it, back to the old spot. When the bee had gone, I put yellow paper where the green had been, and put the green again a foot off.
,, 3. 0 to the honey on the yellow paper. I disturbed the bee, and she at once flew to the honey on the green paper; when she had gone, I put orange paper in the old place, and put the green paper about a foot off.
,, 3.10 to the honey on the green paper

I again gently moved the paper, with the bee on it, to the usual place; and when the bee had gone, put white paper in the old place, and put the green a foot off.

She returned at 3.20 to the honey on the green paper. I again gently moved the green paper, with the bee on it, to the old place; and when she had gone, replaced it by blue paper, putting the green a foot off.

„ 3.30 to the honey on the green paper. I again repeated the same thing, putting yellow instead of blue.

„ 3.40 to the green paper. I now reversed the position of the yellow and green papers; but

„ 3.51 to the green. After this
„ 4. 6
„ 4.15
„ 4.28, when she left off for the day, nor were there any bees still working in the garden. The same afternoon a wasp, which I was observing, remained at work till 6.29 p.m.

August 20.—About noon I brought five bees to some honey at my window. They all soon returned, and numerous friends came with them. One of them I put to some honey on blue paper. She returned as follows, viz. :—

At 12.36	At 2.30
12.42	2.38
12.53	3. 2
1.28	3.10
1.38	3.22
1.49	3.50
2. 2	4. 4
2.11	4.14
2.24	4.23

when I left off watching and shut her out. The longer intervals are due to her having got some honey every now and then on her wings and legs, when she lost a little time in cleaning herself.

August 21.—I opened my window at 6 A.M No bee came till at 7.33 the one above-mentioned came to the honey on blue paper.

I also placed some honey on orange paper about two feet off.

At 7.42 she returned to the honey on blue paper, and again

7.55 she returned to the honey on blue paper.
8. 3 ,, ,,
8.14 ,, ,,

At 8.25 She returned to the honey on blue paper.
 8.35   ,,   ,,
 8.44   ,,   ,,
 8.54   ,,   ,,
 9. 5   ,,   ,,

I then transposed the papers, but not the honey.

At 9.16 she came back to the honey on blue paper. I then transposed the papers again.

At 9.29 she came back to the honey on blue paper. I then transposed them again.

 At 9.39   ,,   ,,   ,,   ,,

At 9.53 she came back to the honey on blue paper. I now put green paper instead of orange, and transposed the places.

At 10.0 she came back to the honey on green paper. I transposed them again.

At 10.8 she came back to the honey on blue paper. I transposed them again.

At 10.21 she came back to the honey on green paper. I now put red paper instead of green, and transposed the places.

At 10.30 she came back to the honey on blue paper I transposed them again.

 At 10.42   ,,   ,,   ,,   ,,
  10.53   ,,   ,,   ,,   ,,
  11. 4   ,,   ,,   ,,   ,,
  11.16   ,,   ,,   ,,   ,,

I now put white paper instead of red, and transposed the places.

At 11.28 she came back to the honey on blue paper
I transposed them again.

   At 11.41   ,,   ,,   ,,   ,,
        11.56   ,,   ,,   ,,   ,,
        12. 8   ,,   ,,   ,,   ,,

At 12.17 she came back to the honey on blue paper. I now put green paper again instead of white, and transposed the places.

At 12.27 she came back to the honey on blue paper. I transposed them again.

   At 12.40   ,,   ,,   ,,   ,,
        12.50   ,,   ,,   ,,   ,,
        1. 0   ,,   ,,   ,,   ,,
        1.13   ,,   ,,   ,,   ,,

At 1.25 she came back to the honey on blue paper, and then to the green. I transposed them again.

At 1.40 she came back to the honey on blue paper. I transposed them again.

At 1.47 she came back to the honey on green paper.

,, 1.57 she came back to the honey on blue paper, and then to the green.

At 2. 6 she came back to the honey on blue paper.

  ,, 2.17   ,,   ,,   ,,   ,,

The following day I accustomed this bee to green paper. She made 63 visits (beginning at 7.47 and ending at 6.44), of which 50 were to honey on green paper.

The following day, August 23, she began work,—

At 7.12 returning to honey on green paper. I then put some on yellow paper about a foot off.

At 7.19 she turned to the honey on green paper. I transposed the colours.

At 7.25 she turned to the honey on green paper. I replaced the yellow paper by orange and transposed the places.

At 7.36 she turned to the honey on green paper. I transposed the colours so that the orange might be on the spot to which the bee was most accustomed.

At 7.44 she turned to the honey on green paper. I now put white instead of orange.

At 7.55 she turned to the honey on green paper. Transposed the papers.

At 8.1 she turned to the honey on green paper. I now put blue paper instead of white.

At 8.12 she turned to the honey on blue paper; but it will be remembered that she had been previously accustomed to come to the blue. I now put red instead of blue.

At 8.23 she turned to the honey on green paper.
 8.25 ,, ,, ,, ,,
 8.47 ,, ,, ,, ,,

I then ceased observing and removed the honey.

Thus the bee which was accustomed to green, returned to that colour when it was removed about a foot, and replaced by yellow, orange, white, and red; but, on the other hand, when it was replaced by blue, she went to the blue. I kept this bee under observation till the 28th, but not with reference to colours.

August 24.—At 7.45 I put another bee to honey on green paper, to which she kept on returning till 9.44. The next day (August 25) she came at 7.38, and I let her come to the green paper till 9. The following morning she returned at 6 A.M., coming back as follows, viz.:—

> At 6.10
> 6.18
> 6.25
> 6.35
> 6.45
> 6.54
> 7. 3
> 7.13

I now put orange in place of green, and put the green a foot off.

At 7.24 she returned to the green. I replaced the paper with the bee on it; and when she had gone I put light blue in place of the green, and again moved the green a foot off.

At 7.36 she returned to the blue. I again replaced the paper with the bee on it; and when she had gone I put yellow in place of the green, and again moved the green a foot off.

At 7.44 she returned to the green. I then did

exactly the same, only putting vermilion in place of the green.

At 7.55 she returned to the green. I then did exactly the same, only putting white in place of green.

At 8. 3 ,, ,, ,, ,,

These observations clearly show that bees possess the power of distinguishing colours.

It remained to determine, if possible, whether they have any preference for one colour over another. M. Bonnier in a recent memoir[1] denies this. He does not question the power of insects to distinguish colours, which he admits that the preceding observations clearly prove, but he maintains that they would not be in any way attracted or guided by the colours of flowers. This he has attempted to demonstrate by experiment. With this view he proceeded as follows: —He took four cubes, 22 centim. by 12 (*i.e.* about 9 inches by $3\frac{1}{2}$), and coloured red, green, yellow, and white, placing them 6 feet apart in a line parallel to and about 60 feet distant from the hives. He then placed on each an equal quantity of honey, and from minute to minute counted the number of bees on each cube. He found that the number of bees on each was approximately equal, and that the honey was removed from each in about twenty minutes. In the experiment he records the bees began to arrive directly the honey was arranged, and in ten minutes there were nearly a hundred bees on each cube. I presume, therefore, that

Les Nectaires.

the bees were previously accustomed to come to the spot in question, expecting to find honey.

I do not think, however, that any conclusive result could be expected from this experiment. In the first place, after the first five minutes there were about thirty bees on each cube, and in less than ten minutes nearly a hundred, and the colour therefore must have been almost covered up. The presence of so many bees would also attract their companions. Moreover, as the honey was all removed in less than twenty minutes, the bees were evidently working against time. They were like the passengers in an express train, turned hurriedly into a refreshment-room; and we cannot expect that they would be much influenced by the colouring of the tablecloth. In fact, the experiment was too hurried, and the test not delicate enough.

Then, again, he omitted blue, which I hope to show is the bee's favourite colour, and his cubes were all coloured. It is true that one was green; but any one may satisfy himself that a piece of green paper on grass is almost as conspicuous as any other colour. To make this experiment complete, M. Bonnier should have placed beside the honey on the coloured cubes a similar supply, without any accompaniment of colour to render it conspicuous.

I could not, therefore, regard these experiments as at all conclusive. The following seem to me a more fair test:--

I took slips of glass of the size generally used for

slides for the microscope, viz. 3 inches by 1, and pasted on them slips of paper coloured respectively blue, green, orange, red, white, and yellow. I then put them on a lawn, in a row, about a foot apart, and on each put a second slip of glass with a drop of honey. I also put with them a slip of plain glass with a similar drop of honey. I had previously trained a marked bee to come to the place for honey. My plan then was, when the bee returned and had sipped about for a quarter of a minute, to remove the honey, when she flew to another slip. This then I took away, when she went to a third; and so on. In this way—as bees generally suck for three or four minutes—I induced her to visit all the drops successively before returning to the nest. When she had gone to the nest I transposed all the upper glasses with the honey, and also moved the coloured glasses. Thus, as the drop of honey was changed each time, and also the position of the coloured glasses, neither of these could influence the selection by the bee.

In recording the results I marked down successively the order in which the bee went to the different coloured glasses. For instance, in the first journey from the nest, as recorded below, the bee lit first on the blue, which accordingly I marked 1; when disturbed from the blue, she flew about a little and then lit on the white; when the white was removed, she settled on the green; and so on successively on the orange, yellow, plain, and red. I repeated the experiment a hundred times, using two different hives—one in Kent and

PREFERENCE FOR PARTICULAR COLOURS. 305

one in Middlesex—and spreading the observations over some time, so as to experiment with different bees, and under varied circumstances. Adding the numbers together, it of course follows that the greater the preference shown for each colour the lower will be the number standing against it.

The following table gives the first day's observations *in extenso* :—

Journeys	Blue	Green	Plain Glass	Orange	Red	White	Yellow
1	1	3	6	4	7	2	5
2	5	4	7	6	1	2	3
3	1	4	7	6	5	3	2
4	2	4	6	7	5	1	3
5	1	4	7	2	6	5	3
6	1	2	3	6	5	4	7
7	2	1	4	7	3	5	6
8	3	4	6	2	7	5	1
9	5	1	7	4	6	3	2
10	1	6	7	5	3	2	4
11	4	6	5	2	7	3	1
	26	39	65	51	55	35	37

In the next series of experiments the bees had been trained for three weeks to come to a particular spot on a large lawn, by placing from time to time honey on a piece of plain glass. This naturally gave the plain glass an advantage; nevertheless, as will be seen, the blue still retained its pre-eminence. It seems hardly necessary to give the observations in detail. The following table shows the general result :—

Series	No. of Exp.	Blue	Green	Orange	Plain	Red	White	Yellow
1st	11	26	39	51	65	55	35	37
2nd, May 30 ...	15	38	57	59	72	66	58	70
3rd, July 2 ...	16	44	76	82	73	53	53	67
4th, ,, 4 .	15	43	61	64	80	66	50	56
5th, ,, 5 ...	10	36	47	39	40	40	36	42
6th, ,, 6 ...	2	2	8	9	10	14	6	7
7th, ,, 20 ...	11	33	39	50	47	49	41	49
8th, ,, 23 ...	10	31	46	48	52	37	35	31
9th, ,, 25 ...	10	22	54	38	52	33	35	46
	100	275	427	440	491	413	349	405

The precautions taken seem to me to have placed the colours on an equal footing; while the number of experiments appears sufficient to give a fair average. It will be observed also that the different series agree well among themselves. The difference between the numbers is certainly striking. Adding together 1, 2, 3, 4, 5, 6, and 7, we get 28 as the total number given by each journey; 100 journeys therefore give, as the table shows, a total of 2,800, which divided by 7 would of course, if no preference were shown, give 400 for each colour. The numbers given, however, are—for the blue only 275, for the white 349, yellow 405, red 413, green 427, orange 440, and plain glass as many as 491.

Another mode of testing the result is to take the per-centage in which the bees went respectively to each colour first, second, third, and so on. It will be observed, for instance, that out of a hundred rounds the bees took blue as one of the first three in 74 cases, and one of the last four only in 26 cases; while, on the contrary, they selected the plain as one of the first three only in 25 cases, and one of the last four in 75 cases.

	Blue	Green	Orange	Plain	Red	White	Yellow
First	31	10	11	5	14	19	9
Second	18	11	13	7	10	21	20
Third	25	12	8	13	16	13	13
Fourth	8	23	15	11	11	12	20
Fifth	11	13	15	19	17	16	10
Sixth	3	15	22	21	18	12	9
Seventh	4	16	16	24	14	7	19
	100	100	100	100	100	100	100

I may add that I was by no means prepared for this result. Müller, in his remarkable volume on Alpine Flowers, states that bees are much more attracted by yellow than by white.[1] In the same work he gives the following table:—

Flowers	In every 100 visits of insects there were			
	Butterflies	Bees	Flies and Gnats	Other insects
3 yellowish-white species	12·8	51·3	15·4	20·5
23 yellow "	47	27·5	28·1	7·2
16 red "	51·4	35·1	9·2	8·2
7 blue "	64·9	26·6	10·7	1·9

This table does not indeed show any absolute preference for one colour rather than another. In the first place, the number of species compared is very different in the case of the different colours; and in

[1] *Alpenblumen*, p. 487.

the second place, the results may of course be due to the taste, quantity, or accessibility of the honey (all of which we know exercise a great influence), rather than by the colour of the flower. Still the table rather seemed to indicate that bees preferred red, white, and yellow, to blue.

I may very likely be asked, if blue is the favourite colour of bees, and if bees have had so much to do with the origin of flowers, how is it that there are so few blue ones? I believe the explanation to be that all blue flowers have descended from ancestors in which the flowers were green; or, to speak more precisely, in which the leaves immediately surrounding the stamens and pistil were green; and that they have passed through stages of white or yellow, and generally red, before becoming blue. That all flowers were originally green and inconspicuous, as those of so many plants are still, has, I think, been shown by recent researches, especially those of Darwin, Müller, and Hildebrand.

But what are the considerations which seem to justify us in concluding that blue flowers were formerly yellow or white? Let us consider some of the orders in which blue flowers occur with others of different colours.

For instance, in the Ranunculaceæ,[1] those with simple open flowers, such as the buttercups and Thalic-

---

[1] I take most of the following facts from Müller's admirable work on Alpine Flowers.

trums, are generally yellow or white. The blue delphiniums and aconites are highly specialised, abnormal forms, and doubtless, therefore, of more recent origin. Among the Caryophyllaceæ the red and purplish species are amongst those with highly specialised flowers, such as *Dianthus* and *Saponaria*, while the simple open flowers, which more nearly represent the ancestral type, such as *Stellaria, Cerastium*, &c., are yellow and white.

Take, again, the Primulaceæ. The open-flowered, honeyless species, such as *Lysimachia* and *Trientalis*, are generally white or yellow; while red, purple, and blue occur principally in the highly specialised species with tubular flowers. The genus *Anagallis* here, however, certainly forms an exception.

Among the violets we find some yellow, some blue species, and Müller considers that the yellow is the original colour. *Viola biflora*, a small, comparatively little specialised fly-flower, is yellow; while the large, long-spurred *V. calcarata*, specially adapted to humble-bees, is blue. In *V. tricolor*, again, the smaller varieties are whitish-yellow; the larger and more highly developed, blue. *Myosotis versicolor* we know is first yellow and then blue; and, according to Müller, one variety of *V. tricolor alpestris* is yellow when it first opens, and gradually becomes more and more blue. In this case the individual flower repeats the phases which in past times the ancestors have passed through.

The only other family I will mention is that of the

Gentians. Here, also, while the well-known deep blue species have long tubular flowers, specially adapted to bees and butterflies, the yellow *Gentiana lutea* has a simple open flower with exposed honey.

Müller and Hildebrand[1] have also pointed out that the blue flowers, which, according to this view, are descended from white or yellow ancestors, passing in many cases through a red stage, frequently vary, as if the colours had not had time to fix themselves, and by atavism assume their original colour. Thus *Aquilegia vulgaris, Ajuga Genevensis, Polygala vulgaris, P. comosa, Salvia pratensis, Myosotis alpestris,* and many other blue flowers, are often reddish or white; *Viola calcarata* is normally blue, but occasionally yellow. On the other hand, flowers which are normally white or yellow, rarely, I might almost say never, vary to blue. Moreover, though it is true that there are comparatively few blue flowers, still, if we consider only those in which the honey is concealed, and which are, as we know, specially suited to and frequented by bees and butterflies, we find a larger proportion. Thus, of 150 flowers with concealed honey observed by Müller in the Swiss Alps,[2] 68 were white or yellow, 52 more or less red, and 30 blue or violet.

However this may be, it seems to me that the preceding experiments show conclusively that bees do prefer one colour to another, and that blue is distinctly their favourite.

[1] *Die Farben der Blüthen,* p. 26.
[2] *Alpenblumen,* p. 492.

CHAPTER XI.

WASPS.

I have also made a few experiments with wasps.

So far as their behaviour, when they have discovered a store of food, is concerned, what has been said with reference to bees would apply in the main to wasps also. I will give some of the details in the Appendix, and here only refer very briefly to some of the experiments.

Experiment 1.—Watched a wasp, which I had accustomed to come to my room for honey, from 9.36 A.M. to 6.25 P.M. She made forty-five visits to the honey, but did not bring a single comrade.

Experiment 2.—The following day this wasp began working—at least, came to my room for the first time at 6.55 A.M., and went on passing backwards and forwards most industriously till 6.17 P.M. She made thirty-eight journeys, and did not bring a single friend.

Experiment 3.—Another wasp was watched from 6.16 A.M. till 6 P.M. She made fifty-one journeys, and during the day five other wasps came to the honey. I do not think she brought them.

Experiment 4.—Another wasp was watched from 10 A.M. to 5.15 P.M.; she made twenty-eight journeys, and brought no friend. This wasp returned the next morning at 6 A.M.

Experiment 5.—A wasp was watched from 11.56 A.M. to 5.36 P.M. She made twenty-three journeys, without bringing a friend

Experiment 6.—Another wasp between 6.40 A.M. and 5.55 P.M. made sixty journeys, without bringing a friend.

Experiment 7.—Another wasp between 7.25 A.M. and 6.43 P.M. made no less than ninety-four visits to the honey, but did not bring a single friend.

Experiment 8.—I watched a wasp on September 19. She passed regularly backwards and forwards between the nest and the honey, but during the whole day only one other wasp came of herself to the honey; this wasp returned on the 20th, but not one other. The 21st was a hot day, and there were many wasps about the house; my honey was regularly visited by the two marked wasps, but during the whole day only five others came to it.

September 22.—Again only one strange wasp came, up to one o'clock.

September 27.—Only one strange wasp came

October 2 *and* 3.—These days were cold; a few marked bees and wasps came to my honey, but no strangers.

October 4.—Two strangers.

October 6.—Only one stranger.

On these days the honey was watched almost without intermission the whole day, and was more or less regularly visited by the marked bees and wasps.

My experiments, then, in opposition to the statements of Huber and Dujardin, serve to show that wasps and bees do not in all cases convey to one another information as to food which they may have discovered, though I do not doubt that they often do so. Of course, when one wasp has discovered and is visiting a supply of syrup, others are apt to come too; but I believe that in many instances they merely follow one another. If they communicated the fact, considerable numbers would at once make their appearance; but I have not often found this to be the case. The frequent and regular visits which my wasps paid to the honey put out for them, prove that it was very much to their taste; yet few others made their appearance.

These and other observations of the same tendency seem to show that, even if wasps have the power of informing one another when they discover a store of good food, at any rate they do not habitually do so.

On the whole, wasps seem to me more clever in finding their way than bees. I tried wasps with the glass mentioned on p. 278, but they had no difficulty in finding their way out.

My wasps, though courageous, were always on the alert, and easily startled. It was, for instance, more difficult to paint them than the bees; nevertheless, though I tried them with a set of tuning-forks covering three octaves, with a shrill whistle, a pipe, a violin, and my own voice, making in each case the loudest and shrillest sounds in my power, I could see no symptoms in any case that they were conscious of the noise.

The following fact struck me as rather remarkable. One of my wasps smeared her wings with syrup, so that she could not fly. When this happened to a bee, it was only necessary to carry her to the alighting-board, when she was soon cleaned by her comrades. But I did not know where this wasp's nest was, and therefore could not pursue a similar course with her. At first, then, I was afraid that she was doomed. I thought, however, that I would wash her, fully expecting, indeed, to terrify her so much that she would not return again. I therefore caught her, put her in a bottle half full of water, and shook her up well till the honey was washed off. I then transferred her to another bottle, and put her in the sun to dry. When she appeared to have recovered I let her out: she at once flew to her nest, and I never expected to see her again. To my surprise, in thirteen minutes she returned as if nothing had happened, and continued her visits to the honey all the afternoon.

This experiment interested me so much that I repeated it with another marked wasp, this time, however, keeping the wasp in the water till she was quite motionless and insensible. When taken out of the water she soon recovered; I fed her; she went quietly away to her nest as usual, and returned after the usual absence. The next morning this wasp was the first to visit the honey.

I was not able to watch any of the above-mentioned wasps for more than a few days, but I kept a specimen of *Polistes gallica* for no less than nine months.

I took her, with her nest, in the Pyrenees, early in May. The nest consisted of about twenty cells, the majority of which contained an egg; but as yet no grubs had been hatched out, and, of course, my wasp was as yet alone in the world.

I had no difficulty in inducing her to feed on my hand; but at first she was shy and nervous. She kept her sting in constant readiness; and once or twice in the train, when the railway officials came for tickets, and I was compelled to hurry her back into her bottle, she stung me slightly—I think, however, entirely from fright.

Gradually she became quite used to me, and when I took her on my hand apparently expected to be fed. She even allowed me to stroke her without any appearance of fear, and for some months I never saw her sting.

When the cold weather came on she fell into a drowsy state, and I began to hope she would hibernate and survive the winter. I kept her in a dark place, but watched her carefully, and fed her if ever she seemed at all restless.

She came out occasionally, and seemed as well as usual till near the end of February, when one day I observed she had nearly lost the use of her antennæ, though the rest of the body was as usual. She would take no food. Next day I tried again to feed her; but the head seemed dead, though she could still move her legs, wings, and abdomen. The following day I offered her food for the last time; but both head and thorax were dead or paralysed; she could but move her tail, a last token, as I could almost fancy, of gratitude and affection. As far as I could judge, her death was quite painless; and she now occupies a place in the British Museum.

Power of distinguishing Colours.

As regards colours, I satisfied myself that wasps are capable of distinguishing colour, though they do not seem so much guided by it as bees are.

July 25.—At 7 A.M. I marked a common worker wasp (*Vespa vulgaris*), and placed her to some honey on a piece of green paper 7 inches by $4\frac{1}{2}$. She worked with great industry. After she had got well used to the green paper I moved it 18 inches off, putting some other honey on blue paper where the green had previously been. She returned to the blue. I then replaced the green paper for an hour, during which she visited it several times, after which I moved it 18 inches, as before, and put brick-red paper in its place. She returned to the brick-red paper. But although this experiment indicates that this wasp was less strongly affected by colours than the bees which I had previously observed, still I satisfied myself that she was not colour-blind.

I moved the green paper slightly and put the honey, which, as before, was on a slip of plain glass, about four feet off. She came back and lit on the green paper, but finding no honey, rose again, and hawked about in search of it. After 90 seconds I put the green paper under the honey, and in 15 seconds she found it. I then, while she was absent at the nest, moved both the honey and the paper about a foot from their previous positions, and placed them about a foot apart. She returned as usual, hovered over the paper, lit on it, rose again, flew about for a few seconds, lit again on the paper, and again rose. After 2 minutes had elapsed I slipped the paper under the honey, when she almost immediately (within 5 seconds) lit on it. It seems obvious, therefore, that she could see green.

I then tried her with red. I placed the honey on brick-red paper, and left her for an hour, from 5 P.M. to 6, to get accustomed to it. During this time she continued her usual visits. I then put the honey and the coloured paper about a foot apart; she returned first to the paper and then to the honey. I then transposed the honey and the paper. This seemed to puzzle her. She returned to the paper, but did not settle. After she had hawked about for 100 seconds I put the honey on the red paper, when she settled on it at once. I then put the paper and the honey again 18 inches apart. As before, she returned first to the paper, but almost immediately went to the honey. In a similar manner I satisfied myself that she could see yellow.

Again, on August 18 I experimented on two wasps, one of which had been coming more or less regularly to some honey on yellow paper for four days, the other for twelve—coming, that is to say, for several days, the whole day long, and on all the others, with two or three exceptions, for at least three hours in the day. Both, therefore, had got well used to the yellow paper. I then put blue paper where the yellow had been, and put the yellow paper with some honey on it about a foot off. Both the wasps returned to the honey on the blue paper. I then moved both the papers about a foot, but so that the blue was somewhat nearer the original position. Both again returned to the blue. I then transposed the colours, and they both returned to the yellow.

Very similar results were given by the wasp watched on September 11. After she had made twenty visits to honey on blue paper, I put it on yellow paper, and moved the blue 12 inches off. She came back to the yellow. I then put vermilion instead of yellow; she came back to the vermilion. I transposed the colours; she came back to the vermilion.

I put white instead of vermilion; she came to the blue.
„ green „ white; she came to the blue.
„ orange „ green; she came to the blue.
I transposed the colours; she returned to the orange.

I put white instead of orange; she came to the white.
„ green „ white; she came to the blue.
„ purple „ green; she came to the purple.
„ orange „ purple; she came to the orange.
„ green „ orange; she came to the green.

I transposed the colours; she came to the blue.
„ „ „ „ green.

So far, therefore, she certainly showed no special predilection for the blue. I then left her the rest of the day to visit the honey on blue paper exclusively. She made fifty-eight visits to it. The following morning I opened my window at 6.15, when she immediately made her appearance.

I let her make ten more visits to the honey on blue paper, moving it about a foot or so backwards and forwards on the table. I then put orange paper instead of the blue, and put the blue about a foot off. She returned to the orange.

I put yellow instead of orange; she came to the yellow.
„ vermilion „ yellow; she came to the vermilion.
„ white „ she came to the white.
„ green „ white; she came to the green.

I transposed the colours; she came to the blue.

I now put vermilion instead of green, and moved both of them a foot, but so that the vermilion was nearest the window, though touching the blue; she came to the vermilion.

Again, September 11, I marked a wasp. She returned to the honey over and over again with her usual assiduity. The following morning I put the honey on green paper; she came backwards and forwards all day. On the 13th I opened my window at 6.8, and she came in immediately. During an hour she made ten journeys. On her leaving the honey for the eleventh time, I placed some honey on vermilion paper where the green had been, and put the honey and the green paper about a foot off.

She came at 7.25 to the vermilion. I then put orange instead of vermilion.

„ 7.34 „ orange. I then put blue instead of orange.

„ 7.40 „ blue. I then put white instead of blue.

„ 7.47 „ white. I then put yellow instead of white.

„ 7.55 „ yellow and then to the green. I transposed the colours.

„ 8. 2 „ green. I then moved both colours about a foot, but so that the yellow was a little nearer to the old place.

She returned at 8.9 to the yellow.

I then removed the yellow paper and honey, and placed the honey which had been on the green paper about a foot from it on the table.

At 8.15 she returned and lit on the green paper, but immediately flew off to the honey. I then transposed the honey and the paper.

At 8.24 she returned and again lit on the paper, but immediately flew off to the honey.

Thus, therefore, though it is clear that wasps can distinguish colours, they appear, as might be expected from other considerations, to be less guided by them than is the case with bees.

I have been much struck by the industry of wasps. They commence work early in the morning, and do not leave off till dusk. I have several times watched a wasp the whole day, and from morning to evening, if not disturbed, they worked without any interval for rest or refreshment.

Being anxious to compare bees and wasps in this respect, on August 6, 1882, I accustomed a wasp and three bees to come to some honey put out for them on two tables, one allotted to the wasp, the other to the bees. The last bee came at 7.15 P.M. The wasp continued working regularly till 7.47, coming at intervals of between six and seven minutes. Next morning, when I went into my study a few minutes after 4 A.M., I found the wasp already at the honey. The first bee came at 5.45, the second at 6.

The wasp occupied about a minute, or even less, in supplying herself with a load of honey, and made during the day, as shown in the Appendix (p. 423), no less than 116 visits to the honey, or 232 journeys between my room and her nest, during which she carried off rather more than sixty-four grains of honey.

It would, however, perhaps be unfair to the bees to regard this as indicating that they are less industrious than wasps. The deficiency may be due to their being more susceptible to cold.

I may add that I then left home for a few days. I covered over the honey, leaving only a small entrance for the wasp. When I returned, on the 12th, I found her still at work, and by herself. It was evident that she had continued her labours, but without bringing any friends to assist her.

Every one has heard of a 'bee-line.' It would be no less correct to talk of a wasp-line. On August 6 I marked a wasp, the nest of which was round the corner of the house, so that her direct way home was not out of the window by which she entered, but in the opposite direction, across the room to a window which was closed. I watched her for some hours, during which time she constantly went to the closed window, and lost much time in buzzing about at it. August 7, I was not able to watch her. August 8 and 9, I watched her from 6.25 A.M., when she made her first visit. She still constantly went to the closed window. August 10 and 11, I was away from home. August 12, she made her first visit at 7.40, and still went to the closed window. August 13, her first visit was at 6.15; she went to the closed window and remained buzzing about there till 7, when I caught her and put her out at the open one by which she always entered. August 15 and 16, she continued to visit the honey, but still, always, even after ten days' experience, continued to go to the closed window, which was in the direct line home ; though, on finding it closed, she returned and went round through the open window by which she entered.

APPENDIX A.

Tables illustrating Experiments on Division of Labour.
(See p. 45.)

TABLE I.—Slaves of Formica fusca. Nest No. 1.

Date	6.30	7	8	9	10	11	12	1	2	3	4	5	6	7	8	9	10
Nov. 20	o	o	o	o	o	o	o	N 3	o Friend marked N 4	o	o	o	o o N 4	o o o o o o o			
21	..	N 3	o	o	o	o	o	o	o	..	..	o	..	o	..	..	
22	o N 4	o	N 3 N 4	o	o	N 3	N 3	o	o	N 4	o	o	o	o	o	o	o
23	o	o	o	o	o	o	o	o	o	..	..	o	..	o	..	o	o
24	o	o	o	o	o	o	o	o	o	..	o	o	o	o	N 3	..	o
25	o	o	o	o	o	o	o	o	o	o	o	o	o	o	o	o	o
26	o	o	o	o	o	o	o	.. N 3 N 5 N 3	o	o	o	o	o	o	o	o o o o	o
27	o	o	N 4 N 3	o N 3	o	o	o	o	o	N 3	o	N 5	.. o N 4	o	o	o o o o N 3 .. o	o o c o
28	o	o	N 4 N 4	o N 4	o	o	o	o	o	o	o	o	o N 3	o o N 4	o o o N 4 N 3		o
29	o	o	o N 3	o	o	o	N 3,4	o	o	N 3	o	o	N 3	o	o o o	.. o	o c o
30	..	o	o	o	o	o	o	o	o	o	o	o	o	o	o	.. o	o
Dec. 1	o	o	o	o	o	o	o	o	..	o	o	o	o	o	o	o	o
2	o	o	o	o	o	o	o	o	o	o	o	o	o	o	o	o	o
3	..	o	o	o	o	o	o	o	o	o	o	o	o	o	o	o	o
4	..	c	o	o	o N 4	o	o	o	o	o	o	o	o	o	o	o	c
5	..	c	o	o	o	o	o	o	o	o	o	o	o	o	o	o	o
6	..	c	o	o	o	o	o	o	o	o	o	o	o	o	o	..	c
7	o	c	o	o	o	o	o	o	o	o	o	o	o	o	o	..	o
8	o	o	o	o	o	o	o	o	o	o	o	o	o	..	o	o	o
9	o	o	o	o	o	o	o	o	o	o	o	o	o	o	..	o	o

DIVISION OF LABOUR.

TABLE I. (continued).

Date	6.30	7	8	9	10	11	12	1	2	3	4	5	6	7	8	9	10
Jan. 8	∶	o	o	N 5	N 5&6	o	o	o	o	N 3	o	o	o	o	N 5	o	N 5
9	o	o	o N 3	o	o	o	o	o	⋮	o	o	o	o	o	o	o	o
10	⋮	o	N 6	o N.3	o	o	⋮	o	N 3 Friend feeling	o	o	o	o	o	o	o	o
11	o	o	o	o	o	o	o	o	o	o	o	o	o	o	o	o	o
12	o	o	o	o	o	o	o	o	o	o	o	o	o	o	o	o	o
13	o	o	o	o	o	o	o	o	o	o	o	o	o	o	o	o	o
14	o	N 3	o	o	o	o	o	o	o	o	o	o	o	o	o	o	o
15	o	o	o	o N7&3	o	o	o Friend marked	o	o	o	o	o N 7	o	o	o	o	o
16	o	o	o N 7	o	o	o N 7	N 7	o	o	o	o	o	o N 6	o	o	o	o
17	⋮	o	o	o	o	o	N 3	o	o	o	o	o	o	o	o	N 6	o
18	o	N 6	o N 6	o	o	o	N 7	o	o	o	o	o	o	o	o	o	o
19	o	o	o	o	o	o	o	o	N 6	N 6	o	o	o	o	o	o	o
20	⋮	N 6	o	N 3	o	⋮	o	o	o	o	o	o	o	o	o	o	o
21	o	o	o	o	o	o	o	o	o	o	o	o	o	o	o	o	o
22	o	o	o	o	o	o	o	o	o	o	N 6	o	o	o	o	o	o
23	o	o	o	o	o	o	o	o	o	o	o	o	o	o	o	o	o
24	o	o	o	o	o	o	o	o	o	o	o	o	o	o	o	o	o
25	o	o	o	o	o	o	o	o	o	o	o	N 3	o	o	o	o	o
26	⋮	o	o	o	o	o	o	o	o	⋮	o	o	o	⋮	o	o	o
27	o	o	o	o	o	N 3	o	o	o	o	o	o	o	o	o	o	o
28	o	o	Friend N 8	o	o	o	o	o	o	o N 8	o	o	o	o	o	o	o
29	o	o	N 8	o	o	o	o	o	o	o	o	o	o	o	o	o	o

DIVISION OF LABOUR. 327

328 DIVISION OF LABOUR.

DIVISION OF LABOUR. 329

TABLE II. (continued).

Date	6.30	7	8	9	10	11	12	1	2	3	4	5	6	7	8	9	10
Jan. 5	o	o	o	o	o	o	o	Impri. N 6	o	o	o	o	o	o	o	o	o
6	o o	o o	o o	o o	o o	o o	o o	o o	o o	o o	: o o	: o o	: o o	o o o o	o o o o	Friend feeding o o o	o o o o
7																	
8	: o						: o o										
9	Friend marked N 9																
10		o	o	o	o	o	: o o	o	o	o	o	N 6	o	N 5&9		Impri. N 5	o
11	:	o o	o o	o o	o o	: o o	o o o o o	o c o o o	o o o o o o	o o o o o o	o o N 9 o o	o o o o N 9 N 9	o N 6 o o o N 9	o o o o N 9 N 9	o o o o o o	o o o N 9 o	o N 9 N 9 o o N 9
12																	
13																	
14																	
15	N 9																
16																	
17	: o	o o	o o	o o	o o	o o	o o	Impri. N 9	o o	o o	o o	o o	o o	o o	Friend marked	o o	o o
18																	
19	o	o	o	o	o	o	N 10	o	N 10	o	N 10	o	o	o	N 10	o	o

DIVISION OF LABOUR. 331

20	o	o	N 11	o	o		o	o	o		o	o	o	o
21	o	o	o	o	o		o	o	o		o	o	o	o
22	o	N 10	o	N 10	N 11		o	o	o		N 12	o	o	o
23	...	...	N 11	Friend marked N 12	...		N 11	N 11	o		o	o	o	o
24	o Imprs.	N 7	o	o	o		N 11 N 11	o	o		o	o	o N 6
25	N 7	o	o	N 11	o		o	o	o		N 5	o N 6	o	o
26	o	o	o	o	N 11		N 11	o	o		o	o	o	o
27	o	o	o	o	o		o	...	o		o	o	o	o

I now put back No. 7.

28	o	o	Friend marked N 11	o	o		o	o	o		o	o	N 6	o
29	o	o	o	o	o		o	o	o		o	o	o	o
30	o	...	o	o	o		N 11	o	o		o	o	N 6	o
31	o	...	o	o	o		o	o	o		o	o	o	o

I put back Nos. 5 and 6.

25	o	o	o	o	o		o	o	o		o	N 6	o	o
26	o	o	o	o	o		N 7	o	o		o	o	o	o
27	o	o	o	N 11	o		o	o	o		o	o	o	o
28	o	o	o	o	o		N 11	o	o		o	N 6	o	o
29	...	o	o	o	N 11		o	o	...		N 11	o	o	o
30											o	o	o	o
31											o	o	o	o

TABLE II. (continued).

Date	6 30	7	8	9	10	11	12	1	2	3	4	5	6	7	8	9	10
Feb. 1	o	o	o	o	o	o	o	o	o	o	o	o	o	o	N6	o	o
2	o	o	o	o	o	o	o	o	o	.	o	o	N11	o	o	o	o
3	.	o	o	o	o	o	o	o	o	.	o	o	N11	.	o	o	o
4	.	o	o	o	o	o	o	o	o	o	o	o	N6	o	o	o	N11
5	o	o	o	o	o	o	o	o	o	o	N6	o	o	o	o	o	o
6	o	N11	N11	o	o	o	o	N11	o	o	o	o	o	N11	o	o	N6
7	o	o	o	o	N11	o	N12	o	N11	o	N11	N12&6	N7	N7	o	o	o
8	.	N11	o	N12	N11	N11	N11	o	o	o	o	o	o	o	N10	o	o
9	o	o	N11	o	N11	o	o	o	o	o	o	o	o	o	N10	o	N11
10	o	o	o	o	o	N11	N12	o	o	o	o	o	o	o	o	o	o
11	o	o	N11	o	o	o	o	o	o	o	o	o	o	o	o	N11	N11&12
12	.	N11	o	N11	o	o	o	o	o	o	o	o	o	o	o	o	o
13	o	o	o	o	o	o	o	o	N11	o	N11	o	o	o	o	N11	o
14	o	o	N11	o	N11	o	o	.	o	o	o	o	o	o	o	N11	o
15	o	o	o	o	N11	o	o	N11	N11	o	o	o	o	o	o	N11	o
16	.	o	N11	o	o	.	.	o	o	o	o	o	o	o	o	N11	o
17	o	o	o	o	N11	o	o	o	o	o	o	o	N11	o	o	o	o
18	.	o	o	o	o	o	o	.	o	o	o	o	N11	o	o	o	o
19	.	o	o	o	o	o	o	o	o	o	o	o	o	o	o	o	o
20	o	o	o	o	o	o	o	o	o	o	o	o	o	o	o	o	o
21	.	o	o	o	o	o	o	.	o	o	.	o	o	o	o	o	o
22	.	.	o	o	o	o	o	o	o	o	o	o	o	N12	o	o	o
23	.	.	o	o	o	o	o	.	o	o	.	o	o	o	o	o	o
24	.	.	.	o	o	o	o	o	o	o	o	o	o	o	o	o	o

APPENDIX B.

The following are the details referred to on p. **122**:--

On August 4, 1875, I separated one of my colonies of *Formica fusca* into two halves, and kept them entirely apart.

On March 15 following I put in a stranger and one of the old companions from the other half of the nest at 7 A.M., and watched them longer than those previously experimented on. The stranger was very soon attacked; the friend seemed quite at home.

June 4, 1876.—8 A.M. Put into the nest a stranger and an old friend. The stranger was at once attacked, and dragged about by one of her antennæ. 9 A.M. The stranger was being attacked; the friend, though not attacked, kept rather away from the other ants. 10.30 A.M. The stranger was attacked, not the friend. 12.30 P.M. ditto, 1 P.M. ditto, 1.30 P.M. ditto, 2 P.M ditto, 2.30 P.M. ditto, 4 P.M. ditto, 4.30 P.M. ditto. 5 P.M The stranger was dragged out of the nest.

June 5.—Put in a stranger and a friend at 9.30. At 10 the stranger was being attacked, not the friend. 10 A.M. ditto, 10.30 A.M. ditto.

At 11 A.M. I put in another stranger and another old friend, when nearly the same thing was repeated. At 11.30 A.M. the stranger was being dragged about by her antennæ; the friend was not attacked. 12 A.M. The stranger was by herself in a corner of the nest. The friend was almost cleaned from the paint by which she was marked. I then put in another friend. At 2 P.M. the stranger was being dragged about by an antenna,

the friend was being cleaned. 2.30 P.M. ditto, 3 ditto.
At 3.30 P.M. the friend was almost clean: the stranger
was being dragged about. 6 P.M. ditto.

June 10.—Repeated the same observation at 10 A.M.,
but transposed the colours by which they were distinguished, so that there might be no question whether
perhaps the difference of treatment was due to the
difference of colouring. At 11 A.M. the friend was all
right, the stranger was being dragged about by an
antenna. 11.30 A.M. the friend all right, the stranger
being dragged about by one leg. 12 A.M. ditto.
12.30 P.M. the friend all right, the stranger being
dragged about by an antenna. 1 P.M. ditto, 2 P.M. ditto,
3 P.M. ditto.

July 3.—Put in a friend and a stranger at 11 A.M.
At 11.30 A.M. the stranger was being dragged about,
the friend was being cleaned. 12 A.M. ditto. 12.30 A.M.
both were now being attacked. 1 P.M. ditto.

This seems to show that some at least of the
ants have forgotten their old friends. Perhaps, however,
these were young ants.

July 16.—Put in two friends at 7.45 A.M. At 8 A.M.
each was being dragged about by an antenna. 8.30 A.M.
one was being dragged about by both antennæ, the
other by both antennæ and one leg. 10 A.M. both were
still attacked, but it is curious that at the same time
others were cleaning off the paint. 12.30 P.M. both
still attacked.

July 17.—Put in a friend at 8.15 A.M. At 8.30 A.M.
they were cleaning her. At 9 A.M. she was almost clean.
9.30 A.M. she seemed quite at home, and had only one
spot of paint on her. 10.20 A.M. ditto.

July 20.—Put in a friend and stranger at 9 A.M.
At 9.30 A.M. the friend seemed all right; the stranger
was in a corner by herself. At 10 A.M. the friend was
being cleaned; the stranger had come out of her corner
and was being fiercely attacked. At 11 A.M. the friend

seemed quite at home and was almost cleaned; the stranger was being dragged about, but was almost cleaned. At 12 A.M. the same thing was going on, and also at 12.30 P.M. At 1.30 P.M. the stranger was still being pulled about; but what struck me as remarkable, the friend also had hold of one of the ants by an antenna. At 2 P.M. the friend was by herself, the stranger was being attacked. At 4 P.M. the friend again had hold of an ant by an antenna; the stranger was being pulled about. At 5 P.M. the friend seemed quite at home in the nest, the stranger was dragged out of the nest. The following morning I was still able to distinguish the friend; she seemed quite at home.

August 5.—Put in a stranger and a friend at 8 A.M. At 8.30 A.M. both were attacked. 9 A.M. ditto, 9.30 A.M. ditto, 10 A.M. ditto, 11 A.M. ditto, 12.30 A.M. ditto.

August 6.—Repeated the experiment at 2 A.M. Both ants hid themselves in corners. At 3.30 A.M. the stranger was being attacked; the friend was in a corner by herself. At 4.30 A.M. both were attacked. 5.30 A.M. ditto.

August 7.—Put in a stranger and a friend at 8.30 A.M. At 8.45 A.M. both were being attacked. 9.30 A.M. ditto, 10 A.M. ditto.

August 8.—Put in a friend at 7 A.M. At 8 A.M. she seemed quite at home with the others. At 9 A.M. they had almost cleaned her. At 9.30 A.M. she seemed quite at home with the others. At 10 A.M. ditto.

August 12.—Put in a friend and a stranger at 7 P.M. Both were immediately attacked. 7.15 A.M. they were being dragged about. 7.45 A.M. ditto, 8 ditto, 8.15 A.M. ditto.

August 13.—Put in a friend at 6.30 A.M. At 7.50 A.M. two attacked her. At 8 A.M. she was being attacked by one ant, but another was cleaning her. 8.15 A.M. ditto. 8.45 A.M. Two were attacking her, one dragging at her by an antenna. 9 A.M. ditto, 9.30

A.M. ditto, 10 A.M. ditto, 10.30 A.M. ditto. Others had almost entirely cleaned off the paint.

At 5 P.M. put a friend and a stranger into the other half of the nest. At 5.15 A.M. the friend seemed quite at home, and had been nearly cleaned; the stranger was being attacked. 5.30 A.M. ditto, 8.15 A.M. ditto. 7.15 A.M. Two of the ants were dragging the stranger out of the nest; the friend had been quite cleaned.

August 14.—At 8.15 A.M. I put an ant from each half of the nest into the other. At 8.30 A.M. one was alone in the corner, the other was being attacked. At 9 A.M. both were being attacked. 9.30 A.M. ditto, 10.30 A.M. ditto; 11.30 A.M. ditto, both, however, being almost cleaned.

August 19.—At 8 A.M. I put into each nest one from the other. The one was received amicably and cleaned, so that after a while I lost sight of her. It was clear that she was received in a friendly manner, because no fighting was going on. At 11 A.M. I put into the same nest another friend: at 11.30 A.M. she was all right, and, being cleaned at 12 A.M., I could no longer distinguish her.

The ant put into the other nest was not so well received. At 9.30 A.M., 11.30 A.M., and 12.30 A.M. she was being dragged about, but she was also being cleaned, and after 12.30 A.M. I lost sight of her. As the paint had been entirely removed, but no ant was being attacked, I have no doubt she was at length recognised as a friend.

August 21.—At 10.15 A.M. I again put into each nest an ant from the other. One was at once cleaned, and I could not find her. I should, however, certainly have seen her if she had been attacked.

The other was at first attacked by one of the ants; but this soon ceased, and they began to clean her. By 11.30 A.M. she was quite at her ease among the other ants, and almost clean. After 12 A.M. I could not see her any more. At 1.40 P.M. I again put into each nest an ant from the other, accompanied, however, in both cases by a stranger. The contrast was most marked, and no one who saw it could have doubted that the friends and strangers were respectively recognised as such, or that they themselves were fully aware of their position.

In the first nest the friend at once joined the other ants, who began to clean her. The stranger ran about in evident alarm, was pursued by the others, and took refuge in a corner. At 2 P.M. the friend was with the other ants, the stranger alone in a corner. At 2.25 P.M. the friend was almost cleaned, and after 2.30 P.M. we could no longer distinguish her: the stranger was still alone. At 3.40 P.M. she came out of her hiding-place and was attacked; after a while she escaped from the nest. At 5.30 P.M. she met one of the ants, and a battle at once began. I separated the combatants and put the stranger back near her own nest, which she at once entered, and where she was soon cleaned by her own friends.

I will now describe the adventures of the other couple. The friend immediately joined the other ants; the stranger was hunted about and soon seized. At 2 P.M. the friend was all right, the stranger being dragged about. At 2.30 P.M. ditto. The stranger was soon afterwards dragged out of the nest. The friend, whom I watched at intervals till 6.30 P.M., continued on the best terms with the others; it was quite clear, therefore, that they did not regard her as a stranger. She herself was not afraid of, and did not avoid them. Still for some time she apparently wished to return to the ants with whom she had recently lived. She came out of the nest, and tried to find her way home. I put her back again, however, and by the evening she seemed to have accustomed herself to the change. I then opened the door of the nest soon after 5 P.M.; but she showed no wish to leave her newly rejoined friends.

September 1.—At 11 A.M. I again put into each half of the nest an ant from the other and a stranger. In the one nest the friend joined the other ants, and seemed quite at home; the stranger, on the contrary, endeavoured to conceal herself, and at length, at 4 in the afternoon, escaped from the nest.

In the other division the friend also appeared quite at home. The stranger, on the contrary, endeavoured to escape, but in the course of the afternoon was attacked and killed.

October 15.—At 8 A.M. I repeated the same experiment. In the first nest, up to 10 A.M., neither ant was attacked; and it is curious that the stranger was licked, and, indeed, almost cleaned. Soon afterwards, however, the ants began to attack her, and at three P.M. she was expelled, the friend, on the contrary, being quite at home. Still the following day, at noon, I found her out of the nest (all the rest being within). This almost looks as if, though safe, she did not feel happy; and I accordingly put her back to her old home, which she at once entered.

In the other division the friend was soon nearly cleaned, and the stranger partly so. The friend seemed quite at home. At 12.30 the stranger was being dragged about by three ants; but after this I lost sight of her.

November 10.—At 11.30 put into one of the divisions a friend and a stranger. At 12 the friend was all right, the stranger was being dragged about by an antenna. From this time till 7 P.M. the stranger was continually being dragged about or held a prisoner, while the friend was quite at home.

November 11.—At 10.15 I put into the other division a friend and a stranger. At 11 the friend was quite at home, and the colour with which I had marked her had been almost cleaned off. The stranger, on the contrary, was being dragged about by two of the ants. After this, however, I could not find her. She had, no doubt, escaped from the nest.

November 12.—The following day, therefore, at 11.30, I again put a friend and a stranger into this division of the nest. The friend seemed quite at home. One of the ants at once seized the stranger by an antenna and began dragging her about. I will give this observation in detail out of my note-book.

At 11.45. The friend is quite at home with the rest; the stranger is being dragged about.

At 12. The friend is all right. Three ants now have hold of the stranger by her legs and an antenna.

At 12.15, 12.30, 12.45, and at 1 the stranger was thus held a prisoner.

At 1.30 one now took hold of the friend, but soon seemed to find out her mistake, and left go again.

At 1.45. The friend is all right. The stranger is being attacked. The friend also has been almost cleaned, while on the stranger the colour has been scarcely touched.

At 2.15. Two ants are licking the friend, while another pair are holding the stranger by her legs.

At 2.30. The friend is now almost clean; so that I could only just perceive any colour. The stranger, on the contrary, is almost as much coloured as ever. She is now near the door, and, I think, would have come out, but two ants met and seized her.

At 3. Two ants are attacking the stranger. The friend was no longer distinguishable from the rest.

At 3.30, 3.40, and 5 the stranger was still held a prisoner.

At 6.0. The stranger now escaped from the nest, and I put her back among her own friends.

December 11.—At 10 A.M. I again put in a friend

and a stranger. The friend was not attacked, and consorted peaceably with the rest. I found her again all right on the following morning. The stranger, on the contrary, was soon attacked and expelled.

December 22.—Repeated the same experiment. The stranger was attacked and driven out of the nest. The friend was received quite amicably.

December 26.—Ditto. The friend was received as usual. I lost sight of the stranger, who probably escaped.

December 31.—Ditto. The stranger, after being dragged about some time in the nest, made her escape. But even outside, having met with an ant accidentally, she was viciously attacked.

January 15, 1877.—Ditto.

January 16.—I put in two friends; but thinking the preceding experiments sufficient, I did not on this occasion add a stranger. Neither of the friends was attacked.

January 19.—Put in two friends at 11 A.M. Neither was attacked, and the following morning they were all right amongst the rest.

January 22.—Put in three friends with the same result.

January 24.—Put in two friends with the same result.

January 26.—Put in three friends with the same result.

February 11.—I put in two friends from the other division at 10 A.M. I looked at 10.15, 10.30, 11, 11.30, 12, 2, 4, and 6 P.M. They were on every occasion quite at home amongst the others.

February 12.—Put in three from the other division at 12. They were quite at home. I looked at them at 12.30, 1, 2, 4, and 6. Only for a minute or two at first one appeared to be threatened.

February 13.—Put in one friend from the other division. The ant was put in at 9.15 A.M., and visited at 9.30, 10, 11, 12, and 1. She was evidently quite at home.

February 15.—Ditto. The ant was put in at 10.15 A.M., and visited at 10.30, 11, 12, 1, 2, 3, and 4. She was not attacked.

February 19.—Ditto. The ant was put in at 10 A.M., and visited at 10.15, 10.30, 11, 12, 1, and 2. She was not attacked.

March 11.—Ditto. Ditto at 9.30 A.M., visited at 10.30, 12.30, 2.30, and 5.30. She was not attacked.

March 12.—Ditto. Ditto at 10 A.M., visited at 12, 2, and 4. She was not attacked.

March 18.—Put in two friends at 1 P.M., visited at 2 and 4. She was not attacked.

April 21.—Put in one friend at 9.30 A.M. At 10 she was all right, also at 12 and 4 P.M. She was not attacked.

April 22.—Put in two friends at 8.30 A.M. Visited them at 9 and 10, when they were almost cleaned. After that I could not find them; but I looked at 2, 4, and 6, and must have seen if they were being attacked.

April 23.—Put in two friends at 12.32. Visited them at 1, 2, 3, 4, and 6 P.M. They were not attacked.

May 13.—Put in two friends and a stranger at 7.45. At 9 the two friends were with the rest. The stranger was in a corner by herself. 11 ditto, 12 ditto. At 1 the friends were all right; the stranger was being attacked. 2, the friends all right; the stranger had been dragged out of the nest. The next morning I looked again; the two friends were all right.

May 14.—Put in the remaining three friends at 10. Visited them at 11, 12, 1, 2, 4, and 6. They were not attacked, and seemed quite at home.

This completed the experiment, which had lasted from August 4, 1875, till May 14, 1877, when the last ones were restored to their friends. In no case was a friend attacked.

The difference of behaviour to friends and strangers was therefore most marked.

The friends were gradually licked clean, and except for a few moments, and that probably by mistake, never attacked. The strangers, on the contrary, were not cleaned, were at once seized, were dragged about for hours with only a few minutes' interval, by one, two, or three assailants.

Though the above experiment seemed to me conclusive, I thought it would be well to repeat it with another nest.

I therefore separated a nest of *Formica fusca* into two portions on October 20, 1876.

On February 25, 1877, at 8 A.M. I put an ant from the smaller lot back among her old companions. At 8.30 she was quite comfortably established among them. At 9 ditto, at 12 ditto, and at 4 ditto.

June 8.—I put two specimens from the smaller lot back as before among their old friends. At 1 they were all right and among the others. At 2 ditto. After this I could not distinguish them amongst the rest; but they were certainly not attacked.

June 9.—Put in two more at the same hour. Up to 3 in the afternoon they were neither of them attacked. On the contrary, two strangers from different nests, which I introduced at the same time, were both very soon attacked.

July 14.—I put in two more of the friends at 10.15. In a few minutes they joined the others, and seemed quite at home. At 11 they were among the others At 12 ditto, and at 1 ditto.

July 21.—At 10.15 I put in two more of the old friends. At 10.30 I looked; neither was being attacked. At 11 ditto, 12 ditto, 2 ditto, 4 ditto, and 6 ditto.

October 7.—At 9.30 I put in two, and watched them carefully till 1. They joined the other ants and were not attacked. I also put in a stranger from another nest. Her behaviour was quite different. She kept away from the rest, running off at once in evident fear, and kept wandering about, seeking to escape. At 10.30 she got out; I put her back, but she soon escaped again. I then put in another stranger. She was almost immediately attacked. In the meantime the old friends were gradually cleaned. At 1.30 they could scarcely be distinguished; they seemed quite at home, while the stranger was being dragged about. After 2 I could no longer distinguish them. They were, however, certainly not attacked. The stranger, on the contrary, was killed and brought out of the nest.

This case, therefore, entirely confirmed the preceding, in which strangers were always attacked; friends were in most cases amicably received, even after more than a year of separation. But while the strangers were invariably attacked and expelled, the friends were not always recognised, at least at first. It seemed as if some of the ants had forgotten them, or perhaps the young ones did not recognise them. Even, however, when the friends were at first attacked, the aggressors soon seemed to discover their mistake, and friends were never ultimately driven out of the nest. This recognition of old friends after a separation of more than a year seems to me very remarkable.

The details are, I fear, tedious, but I have thought them worth giving, because a mere general statement, without particulars, would not give so clear an idea of the result.

APPENDIX C.

The following are the details of the observation recorded on p. 161 : —

At 9.45 I put an ant (N1) to a raisin.
At 9.50 she went to the nest.
9.55 I put another (N2) to the raisin.
10.0 she went to the nest.
10. 0 N1 came back.
10.2 she went to the nest.
10. 7 N1 came back.
10.9 she went to the nest.
10.11 N2 came back.
10.13 she went to the nest.
10.12 N1 came back.
10.14 she went to the nest.
10.13 put another (N3) to the raisin.
10.18 she went to the nest.
10.16 N1 came back.
10.17 she went to the nest.
10.22 N2 came back.
10.24 she went to the nest.

(I here overpainted N2, and she returned no more.)

At 10.24 N1 came back.
10.26 she went to the nest.
10.30 N1 came back.
10.32 she went to the nest.
10.33 N3 came back.
10.35 she went to the nest.

POWER OF COMMUNICATION.

At 10.35 N1 came back. (She met with an accident. At first she seemed a good deal hurt, but gradually recovered.)

At 10.40 N3 came back.
        10.46 she went to the nest.
10.46 a stranger came; I bottled her.
10.47   „   „   „
10.52 N1 came back.
        10.54 she went to the nest.
10.57 N3 came back.
        11.2 she went to the nest.
11. 8 N3 came back.
        11.13 she went to the nest.
11.10 a stranger came; I removed her to a little distance.

At 11.11 a stranger came; marked her N4.

11.16 N3	came.	At 11.18	went.
11.23 N4	„	11.25	„
11.24 N3	„	11.26	„
11.27 N4	„	11.29	„
11.31 N3	„	11.34	„
11.32 N4	„	11.35	„
11.40 N3	„	11.42	„
11.40 N4	„	„	„
11.45 N3	„	11.47	„
„	a stranger came.		
11.48 N1	came.	11.49	„
11.49 N4	„	11.50	„
11.51 N1	„	11.53	„
11.53 N3	„	11.56	„
11.54 N4	„	11.56	„
12. 0 N3	„	12. 2	„
„ N4	„	„	„
„ N1	„	„	„
12. 5 N4	„	12. 7	„
12. 6 N3	„	12. 8	„
12.13 N3	„	12.15	„

At 12.14 N4 came.	12.15 went.	
12.17 a stranger came.		
12.19 N4 came.	12.20	,,
12.20 N3 ,,	12.22	,,
12.21 N1 ,,	12.25	,,
12.25 N4 ,,	12.26	,,
12.27 N3 ,,	12.28	,,
12.30 N4 ,,	12.32	,,
,, a stranger came.		
,, N3 (was disturbed)	12.37	,,
12.38 N4 came.	12.40	,,
12.42 N3 ,,		
12.47 N4 ,,	12.49	,,

Thus during these three hours only six strangers came. The raisin must have seemed almost inexhaustible, and the watched ants in passing and repassing went close to many of their friends; they took no notice of them, however, and did not bring any out of the nest to co-operate with them in securing the food though their regular visits showed how much they appreciated it.

Again (on July 15), an ant belonging to one of my nests of *Formica fusca* was out hunting. At 8.8 I put a spoonful of honey before her. She fed till 8.24, when she returned to the nest. Several others were running about. She returned as follows:—

9.10 to the honey, but was disturbed, ran away, and returned at 10.40.		At 10.53 went back to the nest;		
,,	11.30	,,	11.40	,, ,,
,,	12.5 but was disturbed; she ran away again, but			
,,	1.30	At	1.44	to the nest;
,,	2. 0	,,	2.15	,,
,,	3. 7	,,	3.17	,,
,,	3.34	,,	3.45	,,
,,	4.15	,,	4.23	,,

POWER OF COMMUNICATION.

Returned at 4.52   At 5. 3 went back to the nest.
    ,,   5.56   ,, 6.10   ,,   ,,
    ,,   6.25   ,, 6.45   ,   ,,
    ,,   7.13   ,, 7.18   ,,   ,,
    ,,   7.45   ,, 8. 0   ,,   ,,
    ,,   8.22   ,, 8.32   ,,   ,,
    ,,   9.18   ,, 9.30   ,,   ,,
    ,,   10.10   ,, 10.20   ,,   ,,

During the whole day she brought no friend, and only one other ant found the honey, evidently an independent discovery.

APPENDIX D.

The following are the details referred to on page 164:—

September 24, 1875.—I put out two sets of larvæ; and to one of them I placed two specimens of *Myrmica ruginodis*, which I will call 1 and 2. They returned as follows, carrying off a larva on each journey:—

No. 1.	No. 2.
10.23	
	10.26
10.28	
	10.32
10.34	
	10.37
10.40	
	10.41 bringing a friend.
10.50	
	10.55
	11. 6
	11.16
11.40	
	11.44
11.45	
	11.46 an ant came alone.
	11.56
12. 0	
	12. 6 bringing a friend.
12.11	

POWER OF COMMUNICATION. 349

No. 1.	No. 2.	
	12.15	
12.16		
		12.17 an ant came alone.
	12.22	12.22 ,, ,,
	12.29	
	12.34	
12.36		
	12.40	
		12.45 an ant found the second set of larvæ.
	12.47	
	12.53	
		12.58 two ants found the second set of larvæ.
	12.59	
	1. 5	
1. 6		
		1. 7 an ant found the second set of larvæ.
	1.16	
1.20		
	1.21	
	1.26	
	1.35	
1.42		
	1.47	
	1.54	
1.55 with 2 friends.		
	1.59	
2. 2		
		2. 3 an ant found the larvæ.
	2. 4	
2. 9 with a friend.		
	2.10	
2.16		
	2.18	
2.24		
	2.25	2.25 another ant found the second set of larvæ.
	2.34	

No. 1.	No. 2.
2.36	
	2.41
2.44	
	2.45
	2.50
2.51	
	2.55
	3. 0
3. 1	
	3. 6
3.10	3.10
	3.17
3.18	
	3.22
	3.27
3.28	
	3.36
3.40	
	3.47
3.48	
	3.53
3.55	
	3.59
4. 0	
	4. 7
4. 8	
	4.14
4.16	
	4.20
4.27	
	4.31
4.35	
	4.39 with a **friend.**
4.42	4.42
	4.47
4.53	4.53

POWER OF COMMUNICATION. 351

No. 1.	No. 2.
	4.58
	5. 3
5. 5	
	5. 9
5.17	5.17
5.25	
5.32	
5.40	
5.46	
5.55	
6. 5	
	6. 8
6.11	
	6.16
6.20	

They came no more up to 7.30, when we left off watching. The following morning at 6.5 I found No. 1 wandering about, and evidently on the look-out. I put her to some larvæ; and shortly afterwards No. 2 also found them. Their visits were as follows:—

No. 1	No. 2
6.10	
6.21	
6.36	
	6.42
6.44	
6.52	
7. 1	7. 1
	7. 8
7.11	
	7.12
	7.22
	7.29
	7.35 — 7.30 another ant found the larvæ.
7.40	

No. 1.	No. 2.
7.49	
	7.54
8. 5	
8.13	
8.25	
8.31	
8.39	
8.44	
8.48	

Thus, during this period these two ants carried off respectively 62 and 67 larvæ; 10 strangers found the larvæ, half of them only coming to the set visited by the ants under observation. This seems to show that most of them, at any rate, found the larvæ for themselves.

I will now pass to *Lasius niger*.

September 27, 1875.—At 3.55 P.M. I put an ant of this species to some larvæ. She returned as follows:—

4. 3	5. 5
4.11	5.10
4.21	5.14
4.25	5.18
4.28	5.23
4.31	5.29
4.37	5.40
4.40	5.43
4.44	5.46
4.48	5.50
4.52	5.54
4.56	5.59
5. 0	

when she met with an accident. During this time no other ant came to the larvæ.

On October 1, 1875, at 6.15 A.M., I put three specimens of *Lasius niger* to some larvæ. One did **not** return; the other two behaved as follows:—

No. 1 returned to the larvæ at	No. 2 at	Other ants came at
6.52		
	7.12	
		7.14 to lot 2.
	7.22	
7.30		
	7.32	
7.42	7.42	
		7.45 to lot 3.
	7.50	
7.54		
	8. 0	
8. 1		
8. 6 with a friend.	8. 6	
	8. 9	
8.10		
8.17		
		8.19 to lot 1
		8.23 ,,
8.25		
	8.26	
8.32		
8.36		
		8.37 ,,
	8.38	
8.39		
	8.41	
8.44		
		8.45 ,,

Here I left off watching for half an hour.

9.22		
	9.28	
9.29		
9.35	9.35	

No. 1 returned to the larvæ at	No. 2 at	Other ants came at
9.41		
	9.45	
9.47		
9.50		
	9.52	
9.54 with a friend.		
9.57		
		9.58 to lot 1.
	10. 0	
10. 1		
10. 9		
	10.11	
10.13 with a friend.		
10.16	10.16	
	10.25	
	10.30	
	10.36	
	10.46	
	10.50	
10.55		
	10.58	
11. 0		
	11. 2	
11. 3		
11. 7		
	11. 8	
	11.15	
11.16		
11.19	11.19	
11.23		
	11.25	
11.27		
	11.29 with a friend.	
		11.30
11.33		
	11.35	

POWER OF COMMUNICATION. 355

No. 1 returned to the larvæ at	No. 2 at	Other ants came at
11.37		
11.41		
	11.42	
11.45		11.47 to lot 1.
	11.48	
11.49		
11.53		
	11.59	
12. 1		
12. 4		
12. 8		
	12. 9	
12.11		
		12.14 ,,
12.15	12.15	
12.18		
		12.19 ,,
	12.20	
12.21		
12.25		
	12.29 with a friend.	
12.30		
12.35		
	12.36	
12.39		
12.42		
	12.43	
12.45		
	12.47	
12.48		
12.51		
	12.53	
12.54		
		12.56 ,,
12.57	12.57	
1. 0 with friend.	1. 0	

POWER OF COMMUNICATION.

No. 1 returned to the larvæ at	No. 2 at	Other ants came at
1. 2		
1. 5		
1. 7		
	1. 9	
1.10		
		1.11 to lot 1.
1.13		
	1.14	
1.15		
1.18	1.18	
1.21		
1.24		
1.27		1.27
	1.28	
1.30		
1.33		
	1.35	
1.36		
1.39		
1.42	1.42	
1.45		
		1.46
1.48	1.48	
1.51		
	1.53	
1.57		
	1.59	
2. 1		
2. 4		
	2.15	
2.17		
2.21		
	2.22	
2.25		
2.29		
	2.31	

POWER OF COMMUNICATION. 357

No. 1 returned to the larvae at	No. 2 at	Other ants came at
2.33		
2.37		
	2.39	
2.40		
	2.43	
2.44		
2.47		
	2.49	
2.50		
2.54		
2.57		
3. 0		
	3. 4 with a friend.	
3. 6		
3. 9 with a friend.		
3.12		
3.14		
3.16	3.16	
3.20		
	3.21	
3.23		
3.26	3.26	
3.30	3.30	
3.33	3.33	
3 35	3.35	
3.37		
	3.38	
3.39		
3.41		
3.43		
	3.45	
3.46		
	3.48	
3.49		
3.54		
4. 0		

No. 1 returned to the larvæ at	No. 2 at	Other ants came at
4. 3		
	4. 4	
4. 7		
4.12		
4.15		
4.20		
4.26		
4.29		
4.31		
		4.32
4.34		
4.36		
4.39		
	4.40	
4.42		
		4.43
	4.44	
4.45		
4.49	4.49	
	4.55	
4.56		
	4.58	
4.59		
5. 2	5. 2	
	5. 6 with two friends, after which she came no more.	
5. 7		

The first ant returned at

5.10
5.13
5.15
5.18
5.21
5.25
5.28
5.31
        5.33 to lot 2.

POWER OF COMMUNICATION.

The first ant returned at

5.35	7.28
5.38	7.31
5.41	7.34
5.45	7.38
5.51	7.41
5.54	7.44
6. 0	7.47
6. 4	7.51
6. 7	7.55
6.14	7.59
6.17	8. 2
6.20	8. 5
6.28	8.12
6.31	8.15
6.48	8.18
6.54	8.20
7. 0	8.24
7. 3	8.28
7. 6	8.32
7.11	8.35
7.14	8.38
7.18	8.42
7.21	8.44 another
7.24	8·45 [ant came.
7.25	9.44

We continued to watch till 10.15, but she came no more. She had, however, in the day carried off to the nest no less than 187 larvæ. She brought 5 friends with her; less than 20 other ants came to the larvæ.

October 3.—I put a *Lasius niger* to some larvæ. She returned as follows, viz:—

1.42	2. 4
1.48	2. 8
1.52	2.12 with a stranger.
2. 0	2.15

2.19	4. 7
2.24	4.10
2.27	4.12
2.32	4.15
2.36	4.18
2.40	4.22
2.44	4.25
2.49	4.29
2.57	4.32
3. 1	4.35
3. 4	4.38
3. 7	4.43
3.10	4.46
3.13	4.49
3.15	4.54
3.18	4.57
3.20	5. 0
3.23	5. 3
3.31	5. 6
3.35	5.10
3.38	5.14
3.41	5.18
3.49 with a friend.	5.22
3.51	5.26
3.54	5.29
3.57	She dropped on the floor
4. 1	of my room.
4. 4	

I picked her up; and she returned at

6.40	7. 7 with 3 friends.
6.50	7.11. She now fell into
6.54	some water.
7. 4	

In addition to the above experiments with larvæ, I tried the following with syrup.

April 19.—I put out a little syrup on eleven slips of glass, which I placed on eleven inverted flower-pots on the lawn. At 8.35 a *Lasius niger* found the honey on one of the flower-pots.

8.50	{ she returned to the honey, and at }		9. 5	went back to the nest.	
9.21	,,	,,	9.30	,,	,,
9.42	,,	,,	9.50	,,	,,
10.12	,,	,,	10.21	,,	,,
10.35	,,	,,	10.46	,,	,,
11. 9	,,	,,	11.20	,,	,,
11.45	,,	,,	11.50	,,	,,
11.57	,,	,,	12. 2	,,	,,
12.20	,,	,,	12.30	,,	,,
12.45	,,	,,	12.53	,,	,,
1. 8	,,	,,	1.18	,,	,,
1.34	,,	,,	1.43	,,	,,
1.57	,,	,,	2. 7	,,	,,
2.28	,,	,,	2.33	,,	,,
2.49	,,	,,	2.53	,,	,,
2.59	,,	,,	3. 2	,,	,,
3. 9	,,	,,	3.11	,,	,,
3.29	,,	,,	3.30	,,	,,
3.59	,,	,,	4. 8	,,	,,

After which I watched till 6 P.M.; but she did not return again to the honey. During the above time eight ants came to the same honey, and twenty-one to the other ten deposits.

On July 11 I put one of my specimens of *Lasius niger* to some honey at 7.10. She fed till 7.25, when she returned to the nest.

At 7.32	she returned.		At 7.36	another ant came,	
7.47	,,		7.50	[whom I imprisoned.	
8. 0	,,		8.11	,,	,,
8.18	,,				

At 8.36 she returned.
8.59 ,,
9·17 ,,
9.38 ,,
9.53 ,,
10.10 ,,
10.27 ,,
10.44 ,,
11. 6 ,,
11.16 ,,
11.38 ,,
12. 0 ,,
12.36 ,,   At 12.45 another ant came,
              [whom I imprisoned.
12.56 ,,
1.21 ,,
1.44 ,,
2.10 ,,
2.21 ,,
2.29 ,,
2.50 ,,   2.51   ,,
3. 5 ,,

After this she did not come back any more up to 8 P.M.

April 25 was a beautiful day. At 9 A.M. I put some syrup in the same way on five inverted flower-pots, and at

9.10 put an ant to one of the deposits of syrup. At
9.34 another ant came to the same syrup. This one I will call No 2. At
9.40 No. 1 returned.
10.45 No. 2 ,,   At 11 one came to the same honey; this one I will call No. 3.
11. 7 No. 1 ,,   but did not come back any more
12.31 No. 2 ,,   and at 12.47 went

POWER OF COMMUNICATION. 363

1.15 No. 3 returned, and at 1.25 went.
1.22 No. 2 　　,,　　　,,　　1.48 ,,
1.54 No. 3 　　,,　　　,,　　2. 3 ,,
2.18 No. 2 　　,,　　　,,　　2.30 ,,
2.35 No. 3 　　,,　　　,,　　2.36 ,,
2.56 No. 2 　　,,　　　,,　　3. 1 ,,
3.24 No. 2 returned.
4.19 No. 2 　,,

After which I went on watching till 7, but none of these three returned. During the day 7 ants came to this honey, and 27 to the other four deposits. Here, therefore, it is evident that the three watched ants did not communicate, at any rate, any exact information to their friends.

June 27, 1875.—I placed four inverted glasses (tumblers) on the grass, and on the top of each placed a little honey. I then, at 8 o'clock, put two ants, belonging to *F. nigra*, to the honey on one of the glasses.

At 8.25 No. 1 came back, and at 8.45 she returned to the nest, but did not come to the honey any more.

At 9.5 No. 2 came out and wandered about; I put her to the honey again; she fed and at 9.22 returned to the nest.

At 9.28 {she returned to the honey, and at} 9.45 {went back to the nest.}
10.42 　,,　　,,　　10.50 　,,
10.58 　,,　　,,　　11.10 　,,
11.21 　,,　　,,　　11.39 　,,
12.45 　,,　　,,　　12.59 　,,
1.40 　,,　　,,

I continued to watch till 7 P.M., but neither of them returned any more.

August 7, 1875.—I put out four small deposits of honey (which I continually renewed) on slips of glass placed on square pieces of wood, and put an ant (*L. niger*) to one of them at 9.20. She fed and went away.

At 9.35 she returned, and fed till 9.43
   10.14   ,,   ,,   10.17
   10.25   ,,   ,,   10.27
   10.37   ,,   ,,   10.40

This time a friend came with her.

At 10.47 she returned, and fed till 10.53
   11. 0   ,,   ,,   11.14
   11.35   ,,   ,,   11.40
   11.52   ,,   ,,   11.55
   12.13   ,,   ,,   12.16
   1. 0   ,,   ,,   1. 5
   1.15   ,,   ,,   1.18
   1.26   ,,   ,,   1.29
   1.45   ,,   ,,   1.48
   1.58   ,,   ,,   2. 1
   2. 9   ,,   ,,   2.14
   2.20   ,,   ,,   2.21   She was dis-[turbed.
   2.25   ,,   ,,   2.30
   2.37   ,,   ,,   2.40
   3. 2   ,,   ,,   3. 8
   3.16   ,,   ,,   3.20
   3.39   ,,   ,,   3.41
   3.58   ,,   ,,   4. 2
   4.13   ,,   ,,   4.20
   4.29   ,,   ,,   4.36

At this time there was a shower of rain, so I removed the honey for half an hour.

At 5. 2 she returned, and fed till 5.10
   5.20   ,,   ,,   5.25
   5.33   ,,   ,,   5.37
   5.42   ,,   ,,   5.45
   5.50   ,,   ,,   5.52
   5.58   ,,   ,,   6. 6
   6.15   ,,   ,,   6.18
   6.21   ,,   ,,   6.23

At 6.25 she returned, and fed till 6.27
   6.32   ,,   ,,   6.35
   6.40   ,,   ,,   6.44
   6.49   ,,   ,,   6.53
   7.15   ,,   ,,   7.20
   7.25   ,,   ,,   7.27
   7.30   ,,   ,,   7.33
   7.36   ,,   ,,   7.37

During the whole of this time only three other ants came to the honey.

On January 3, 1875, I placed some larvæ in three small porcelain saucers in a box 7 inches square attached to one of my frame-nests of *Lasius flavus* (Pl. I. Fig. 2). The saucers were in a row 6 inches from the entrance to the frame, and 1½ inch apart from one another.

   At 1.10 an ant came to the larvæ in the cup which I will call No. 1, took a larva, and returned to the nest.
   1.24 she returned and took another.
   1.45   ,,   ,,
   2.10 she went to the further saucer, No. 3. I took her up and put her to No. 1. She took a larva and returned.
   2.24 she returned to cup No. 3. As there were only two larvæ in this cup, I left her alone. She took one and returned.
   2.31 she returned to cup No. 3 and took the last larva.
   2.40 she came back to cup No. 3 and searched diligently, went away and wandered about for two minutes, then returned for another look, and at length at 2.50 went to cup No. 1 and took a larva.
   3. 0 came to cup 1 and took a larva.
   3. 7   ,,   ,,

3.15 came to cup 1 and took a larva, first, however, going and examining cup 3 again.
3.18 came to cup 3, then went to cup 2 and took a larva.
3.30 came to cup 3, then went to cup 2 and took a larva.
3.43 came to cup 3, then went to cup 2 and took a larva.
5.53 came to cup 3, but did not climb up it, then went to cup 2 and took a larva, which she either dropped or handed over to another ant; for without returning to the nest, at 3.55 she returned to the empty cup, and then to cup 2, where she took the last larva, so that two cups are now empty.
4. 3 she came to cup 3, then to cup 2, and lastly to cup 1, when she took a larva.
4.15 came to cup 1 and took a larva.
4.22 ,, ,,
4.38 ,, ,,
5. 0 came to cup 3, then to cup 2, and lastly to cup 1, when she took a larva.
5.19 came to cup 1 and took a larva.
5.50 ,, 2, and then to cup 1 and took a larva.
6.20 ,, 1 and took the last larva.

I now put about 80 larvæ into cup 3.

It is remarkable that during all this time she never came straight to the cups, but took a roundabout and apparently irresolute course.

At 7.4 she came to cup 1 and then to cup 3, and then home. There were at least a dozen ants exploring in the box; but she did not send any of them to the larvæ.

At 7.30 she returned to cup 3 and took a larva.

I now left off watching for an hour. On my return at 8.30 she was just carrying off a larva.

8.40 she came back to cup 3 and took a larva.

8.55 she came to cup 1, then to cup 3 and took a larva.

9.12 ,, ,, ,, ,,
9.30 ,, 3 ,, ,,
9.52 ,, ,, ,, ,,
10.14 ,, 1 ,, ,,

10.26 she went and examined cup 2, then to cup 3 and took a larva.

At 10.45 she came to cup 3, and I went to bed. At 7 o'clock the next morning the larvæ were all removed. In watching this ant I was much struck by the difficulty she seemed to experience in finding her way. She wandered about at times most irresolutely, and, instead of coming straight across from the door of the frame to the cups, kept along the side of the box; so that in coming to cup 3 she went twice as far as she need have done. Again, it is remarkable that she should have kept on visiting the empty cups time after time. I watched for this ant carefully on the following day; but she did not come out at all.

During the time she was under observation, from 1 till 10.45, though there were always ants roaming about, few climbed up the walls of the cup. Five found their way into the (empty) cup 1, and one only to cup 3. It is clear, therefore, that the ant under observation did not communicate her discovery of larvæ to her friends.

The following day I watched again, having, at 7 A.M., put larvæ into one of the porcelain cups arranged as before. No ants found them for several hours.

At 11.37 one came and took a larva.
,, 11.50 she returned and took a larva.
,, 11.59 ,, ,,
,, 12. 9 ,, ,,
,, 12.16 ,, ,,

At 12.21 she returned and took a **larva**.
,, 12.26 ,, ,,
,, 12.32 ,, ,,
,, 12.37 ,, ,,
,, 12.41 ,, ,,
,, 12.45 ,, ,,
,, 12.50 ,, ,,
,, 12.57 ,, ,,
,, 1. 5 ,, ,,
,, 1.11 ,, ,,
,, 1.21 ,, ,,
,, 1.35 ,, ,,
,, 1.40 ,, ,,
,, 1.44 ,, ,,
,, 1.52 ,, ,,
,, 1.56 ,, ,,
,, 2. 2 ,, ,,
,, 2.10 ,, ,,
,, 2.17 ,, ,,
,, 2.24 ,, ,,
,, 2.30 ,, ,,
,, 2.36 ,, ,,
,, 2.43 ,, ,,
,, 2.48 ,, ,,
,, 2.54 ,, ,,
,, 2.59 ,, ,,
,, 3. 3 ,, ,,
,, 3.10 ,, ,,
,, 3.14 ,, ,,
,, 3.19 ,, ,,
,, 3.34 ,, ,,
,, 3.39 ,, ,,
,, 3.47 ,, ,,
,, 3.56 ,, ,,
,, 4. 7 ,, ,,
,, 4.13 ,, ,,
,, 4.20 ,, ,,

CO-OPERATION.

At 4.28 she returned and took a larva.
,, 4.39 ,, ,,
,, 4.44 ,, ,,
,, 4.50 ,, ,,
,, 4.55 ,, ,,
,, 5. 1 ,, ,,
,, 5. 7 ,, ,,
,, 5.17 ,, ,,
,, 5.23 ,, ,,
,, 5.28 ,, ,,
,, 5.40 ,, ,,
,, 5.45 ,, ,,
,, 5.59 ,, ,,
,, 6. 9 ,, ,,
,, 6.13 ,, ,,
,, 6.35 ,, ,,
,, 6.40 ,, ,,
,, 6.46 ,, ,,
,, 6.51 ,, ,,
,, 6.58 ,, ,,
,, 7. 2 ,, ,,
,, 7. 8 ,, ,,
,, 7.12 ,, ,,
,, 7.16 ,, ,,
,, 7.21 ,, ,,
,, 7.26 ,, ,,
,, 7.39 ,, ,,
,, 7.44 ,, ,,
,, 7.53 ,, ,,
,, 7.57 ,, ,,
,, 8. 3 ,, ,,
,, 8. 8 ,, ,,
,, 8.13 ,, ,,
,, 8.20 ,, ,,
,, 8.26 ,, ,,
,, 8.31 ,, ,,
,, 8.38 ,, ,,

At 8.45 she returned and took a **larva.**
,, 8.50 ,, ,,
,, 8.55 ,, ,,
,, 9. 2 ,, ,,
,, 9.11 ,, ,,
,, 9.19 ,, ,,
,, 9.25 ,, ,,
,, 9.33 ,, ,,
,, 9.40 ,, ,,
,, 9.46 ,, ,,
,, 9.52 ,, ,,
,, 10.32 ,, ,,
,, 10.39 ,, ,,
,, 10.49 ,, ,,
,, 10.54 ,, ,,
,, 11. 1 ,, ,,

At this time I went to bed. There were still about twenty-five larvæ in the cup, which had all been removed when I looked at 6.15 the next morning During the whole time she was under observation, only two other ants found their way to the cup, though there were some wandering about in the box all day. Towards evening, however, they went into the nest, and for some hours my ant was the only one out. It will be observed that she returned at shorter intervals than the previous ones. This was partly because she had a shorter distance to go, and partly because she was not bewildered by three cups, like the preceding. I had placed a bit of wood to facilitate her ascent into the cup. This she made use of, but instead of going the shortest way to the cup, she followed the side of the box, partly, perhaps, because the floor was covered with a plate of porcelain. This, however, would not account for the fact that at first she invariably went beyond the cup, and even past the second cup; gradually, however, this circuit became smaller and smaller; but to the last she went round the outside of cup 1, instead of going straight to the spot where I had placed the bit of wood.

On January 9 again I watched her under similar circumstances. From 9.35 to 1.40 she made 55 journeys to and fro, carrying off a larva each time; but during this period only one other ant found the larvæ.

In the afternoon of the same day I watched the ant which had been under observation on the 3rd Jan. From 3.27 to 9.30 she made forty-two visits, during which time only four other ants came to the larvæ.

On January 10 I watched the same ant as on the 4th. Between 11 A.M. and 10 P.M. she made no less than ninety-two visits; and during the whole time only one strange ant came to the larvæ.

On January 18 I put out some more larvæ in the small porcelain cups. Between 8 and 9 both these ants found them, and kept on coming all day up to 7 P.M., when I left off observing. There were a good many ants wandering about in the box; but up to 4 o'clock only four came to the larvæ. Two of them I imprisoned as usual; but two (which came at 4.30 and 4.36) I marked. These went on working quietly with the first two till I left off observing at 7 P.M.; and during this latter time only three other ants found the larvæ.

On January 31 I watched another specimen. At 9.14 I put her into a small cup containing a number of larvæ. She worked continuously till half-past seven in the evening, when I left off watching. During that time she had made more than ninety journeys, carrying each time a larva to the nest. During the whole time not a single other ant came to the larvæ.

Again, on February 7, I watched two ants in the same manner. At 7 A.M. I put some larvæ in the small china cups. Up to 8 no ants had come to them. Soon after 8 I put two marked ants, neither of them being the same as these whose movements are above recorded.

They were then watched until a quarter to eight in the evening, during which time one of them had made twenty-six journeys, carrying off a larva each time; the other forty-two. During this period of about eleven hours, two other ants had come to the cup at which these were working, and the same number to one of the other cups.

None of these ants, therefore, though they had found a large number of larvæ, more than they could carry in a whole day, summoned any other to their assistance.

Again, February 7, 1875, I put some larvæ in three porcelain cups in the feeding-box of a frame containing a nest of *Lasius flavus*, about six inches from the entrance of the frame, and put, at 8 and 8.29 A.M. respectively two ants to the larvæ in the left-hand cup. They each carried off a larva and returned as follows:—

	No. 1.	No. 2.		
At	8.35	—	returned again and took another.
	9. 0	—	,,	,,
		9. 7	,,	,,
		9.20	,,	,,
	9.30	—	,,	,,
		9.43	,,	,,
	9.54	—	,,	,,
		9.56	,,	,,
		10.20	,,	,,
	10.25	—	,,	,,

At 10.43 another ant came to the larvæ in the right-hand cup. I imprisoned her.

	No. 1.	No. 2.		
At		11. 0	returned again and took another.
	11. 1	—	,,	,,
		11. 9	,,	,,
	11.15	—	,,	,,
		11.20	,,	,,
		11.29	,,	,,

EXPERIMENTS ON CO-OPERATION. 373

At 11.37 — returned again and took another.
        11.40   ,,   ,,
        11.52   ,,   ,,

At 12.2 another ant came to the larvæ in the left-hand cup. I imprisoned her.

At 12. 3 — returned again and took another
        12.15   ,,   ,,
        12.30   ,,   ,,
12.37 —   ,,   ,,
        12.41   ,,   ,,
        12.50   ,,   ,,
        12.58   ,,   ,,
1. 0 —   ,,   ,,
        1. 7   ,,   ,,
1.12 —   ,,   ,,
        1.16   ,,   ,,
        1.28   ,,   ,,
1.32 —   ,,   ,,
        1.35   ,,   ,,
        1.44   ,,   ,,
1.50 —   ,,   ,,
        1.55   ,,   ,,
        2. 6   ,,   ,,
2. 9 —   ,,   ,,
        2.17   ,,   ,,
        2.29   ,,   ,,
2.39 —   ,,   ,,
        2.42   ,,   ,,
2.49 2.49   ,,   ,,
3. 0 —   ,,   ,,
        3. 3   ,,   ,,

At 3.10 another ant came to the left-hand cup. I imprisoned her.

At   3.14 returned again and took another.
  3.15 —   ,,   ,,
        3.24   ,,   ,,

At 3.31 — returned again and took another.
3.34 " "
3.36 — " "

At 4.10 another ant came to the middle cup. I imprisoned her.

At 4.45 — returned again and took another.
5.50 " "
6. 2   6. 2 " "
6.17 " "
6.26 " "
6.46 " "
6.52 " "
7. 4   — " "
7. 7 " "
7.13 " "
7.18 " "
7.48   7.48 " "

After this they were not watched any more. It will be observed that the second ant made many more visits than the first—namely, forty-two in about eleven hours, as against twenty-six in eleven hours and a half. During this time two ants came to the larvæ in the cup they were visiting, and three to the other two cups.

The following case is still more striking. On July 11, 1875, at 11 A.M., I put a *Lasius flavus* to some pupæ of the same species, but from a different nest. She made eighty-six journeys, each time carrying off a pupa with the following intervals. Commencing—

At 11. 0   At 11.29 again.
11. 5 she returned.   11.49 "
11. 9 returned again.   11.55 "
11.16 again.   12. 0 "
11.20 "   12. 5 "
11.24 "   12.16 "
11.26 "   12.30 "

EXPERIMENTS ON CO-OPERATION. 375

At 12.40 again.	At 3.40 again.
12.44 ,,	3.47 ,,
12.50 ,,	3.53 ,,
1. 1 ,,	3.57 ,,
1.10 ,,	4. 0 ,,
1.19 ,,	4. 3 ,,
1.27 ,,	4. 5 ,,
1.33 ,,	4. 8 ,,
1.43 ,,	4.12 ,,
1.49 ,,	4.15 ,,
1.52 ,,	4.18 ,,
1.56 ,,	4.20 ,,
2. 2 ,,	4.23 ,,
2.10 ,,	4.26 ,,
2.17 ,,	4.30 ,,
2.25 ,,	4.33 ,,
2.29 ,,	4.40 ,,
2.32 ,,	4.43 ,,
2.35 ,,	4.45 ,,
2.37 ,,	4.49 ,,
2.40 ,,	4.53 ,,
2.43 ,,	4.55 ,,
2.47 ,,	4.58 ,,
2.53 ,,	5. 3 ,,
2.56 ,,	5. 7 ,,
2.59 ,,	5.12 ,,
3. 2 ,,	5.19 ,,
3. 7 ,,	5.22 ,,
3.10 ,,	5.25 ,,
3.13 ,,	5.28 ,,
3.16 ,,	5.32 ,,
3.20 ,,	5.35 ,,
3.25 ,,	5.39 ,,
3.33 ,,	5.50 ,,
3.35 ,,	7. 5 ,,
3.38 ,,	7.12 ,,

After which she did not come again till 8, when we left off watching. During the whole of this time she did not bring a single ant to help her. Surely it would have been in many respects desirable to do so. It will be seen that some of the pupæ remained lying about and exposed to many dangers from 11 A.M. till 7 P.M.; and when she left off working at that time, there were still a number of the pupæ unsecured; and yet, though she had taken so much pains herself, she did not bring or send others to assist her in her efforts or to complete her work.

I have given the above cases at length, though I fear they may appear tedious and prolix, because they surprised me much.

No doubt it more frequently happens that if an ant or a bee discovers a store of food, others soon find their way to it, and I have been anxious to ascertain in what manner this is effected. Some have regarded the fact as a proof of the power of communication; others, on the contrary, have denied that it indicated any such power. Ants, they said, being social animals, naturally accompany one another; moreover, seeing a companion coming home time after time with a larva, they would naturally conclude that they also would find larvæ in the same spot. It seemed to me that it would be very interesting to determine whether the ants in question were brought to the larvæ, or whether they came casually. I thought therefore that the following experiment might throw some light on the question, viz.: to place several small quantities of honey in similar situations, then to bring an ant to one of them, and subsequently to register the number of ants visiting each of the parcels of honey, of course imprisoning for the time every ant which found her way to the honey except the first. If, then, many more came to the honey which had been shown to the first ant than to the other parcels, this would be in favour of their possessing the power of communicating facts to one another, though it might be said they came by scent. Accordingly, on July 13, 1874, at 3 P.M., I took a piece of cork about 8 inches long and 4 inches wide, and stuck into it seventeen pins, on three of which I put pieces of card with a little honey. Up to 5.15 no ant had been up any of these pins. I then put an ant (*Lasius niger*) to the honey on one of the bits of card. She seemed to enjoy it, and fed for about five minutes, after which she went away. At 5.30 she returned, but went up six pins which had no honey on them. I then put her on to the card. In the mean time twelve other ants went up wrong pins and two up to the honey; these I imprisoned for the afternoon. At 5.46 my first ant went away. From that time to 6 o'clock seven ants came, but the first did not return. One of the seven went up a wrong pin, but seemed surprised, came down, and immediately went up to the right one. The other six went straight up the right pin to the honey. Up to 7 o'clock twelve more ants went up pins—eight right, and four wrong. At 7 two more went wrong. Then my first ant returned, bringing three friends with her; and they all went straight to the honey. At 7.11 she went home: on her way to the nest she met and accosted two ants, both of which then came straight to the right pin and up it to the honey. Up to 7.20 seven more ants came and climbed up pins—six right, and one wrong. At 7.22 my first ant came back with five friends; at 7.30 she went away again, returning at 7.45 with no less than twenty companions. During this experiment I imprisoned every ant that found her way up to the honey. Thus, while there were seventeen pins, and consequently sixteen chances to one, yet between 5.45 and 7.45 twenty-seven ants came, not counting those which were brought by the original ant; and out of these twenty-seven, nineteen went straight up the right pin. Again, on the 15th July, at 2.30, I put out the same piece of cork with ten pins, each with a piece of card and one with honey. At 4.40 I put an ant to the honey; she fed comfortably, and went away at 4.44.

At 4.45 she returned, and at 5. 5 went away again.
„ 5.40   „   „   5.55   „
„ 6.13   „   and again at 6.25 and 6.59.

There were a good many other ants about, which, up to this time, went up the pins indiscriminately.

At 7.15 an ant came and went up the right pin, and another at 7.18. At 7.26 the first ant came back with a friend, and both went up the right pin. At 7.28 another came straight to the honey.

At 7.30 one went up a wrong pin.
„ 7.31 one came to the right pin.
„ 7.36   „   „   with the first ant.
„ 7.39   „   „
„ 7.40   „   „
„ 7.41   „   „
„ 7.43   „   „
„ 7.45   „   „
„ 7.46   „   „
           „   „   wrong pin.
           „   „   „
„ 7.47 two   „   „
„ 7.48 one   „   right pin.
    „   the first ant came back.
„ 7.49 another ant came to the right pin.
„ 7.50   „   „   wrong „
„ 7.51   „   „   right „
    „   three ants   „   wrong „
„ 7.52 one ant   „   right „
„ 7.55   „   „   wrong „
    „   „   „   right „
„ 7.57   „   „   wrong „

POWER OF COMMUNICATION. 379

At 7.58 one ant came to the right pin.
„ 7.59 „ „ wrong „

Thus after 7 o'clock twenty-nine ants came; and though there were ten pins, seventeen of them went straight to the right pin.

On the 16th July I did the same again. At 6.25 I put an ant to the honey; at 6.47 she went.

At 6.49 an ant came to the right pin.
„ 6.50 another „ „
„ 6.55 „ „ „
„ 6.56 „ „ wrong pin, and then to the right one.
„ 6.58 „ „ right pin.
„ 7. 0 „ „ „
„ 7. 5 the first ant came back, and remained at the honey till 7.11.
„ 7. 5 another came to the right pin; but she was with the first.
„ 7. 6 another ant came to the right pin.
„ 7. 6 „ „ „
„ 7.12 „ „ „
„ 7.13 „ „ „

These two ants were met by the first one, which crossed antennæ with them, when they came straight to the honey.

At 7.14 another ant came straight to the honey.
„ 7.21 the first ant returned; at 7.26 she left.
„ 7.24 another ant came, but went first to a wrong pin, and then on to the right one.
„ an ant came to wrong pin.
„ „ „ „
„ „ „ „
„ 7.34 „ „ „
„ 7.35 „ „ „
„ 7.38 the first came back, at 7.45 went away again.

At 7.42 an ant went to a wrong pin.
„ 7.47   „   „   „
„ 7.48   „   „   „
„ 7.49   „   „   „
„ 7.52   „   „   the right pin.
„ 7.55 the first ant returned, and at 7.56 went away again.
„ 7.57 an ant went to wrong pin.
„ 7.58   „   right „
„ 8. 0   „   wrong „
„   „   right „
„ 8. 1   „   wrong „

After this, for an hour no more ants came. On this occasion, therefore, while there were ten pins, out of thirty ants, sixteen came to the right one, while fourteen went to one or other of the nine wrong ones.

July 18.—I put out the boards as before at 4 o'clock. Up to 4.25 no ant came. I then put one (No. 1) to the honey; she fed for a few minutes, and went away at 4.31.

At 4.35 she came back with four friends, and went nearly straight to the honey. At 4.42 she went away, but came back almost directly, fed, and went away again.

At 4.57 she returned, and at 5.8 went away again.
„ 4.45 an ant came to wrong pin.
„ 4.47   „   „
„ 4.49   „   „
„ 4.50   „   right pin.
„ 4.52   „   „
„ 4.55   „   wrong pin.
„ 4.56   „   right pin. This ant (No. 2) I allowed to return to the nest, which she did at 5.23.
„ 5. 6   „   right pin.
„ 5.11   „   wrong pin.

POWER OF COMMUNICATION.

At 5.12 an ant came to right pin.
    I changed the pin.
„ 5.16 an ant came to the pin which I had put in
      the same place.
„ „   „ right pin.
„ 5.19   „   „
„ 5.20 two ants   „ with No. 2.
 „ ant No. 1 „ and went at 5.25.
„ 5.25 an ant   „ This ant had been
           spoken to by No. 2.
„ 5.26 another ant „
„ 5.35 „   „
„ 5.37 „   „
„ 5.40 „   „
„ 5.41 ant No. 1   „ and went at 5.49
„ 5.45 another ant „
„ 5.50 „   „
„ 5.51 ant No. 1 came back, and 5.54 went.
„ 5.58 two ants came to the right pin.
„ 5.59 another ant „ „
 „ „ came to a wrong pin.
    I changed the pin again.
„ 6.49 an ant came to the pin which I had put in
      the same place.
„ 7. 1 another ant came to the right pin.
„ 7.20   „   „
„ 7.33   „   „
„ 7.46 ant No. 1 returned, 7.55 went.

Thus during this time, from 4.50 until 7.50, twenty-nine ants came, twenty-six went to the right pin, while only three went up any of the nine wrong ones. Moreover, out of these twenty-six, only four were distinctly brought by the two ants which I had shown the honey.

On the 19th I tried a similar experiment. The marked ants frequently brought friends with them; but, without counting these, from 3.20 to 8 o'clock, out of forty-five ants, twenty-nine went up the right pin, while sixteen went up the nine wrong ones.

Thus on

July 13,	out of	27	ants,	19	went right and	8	wrong.
,, 15	,,	29	,,	17	,,	12	,,
,, 16	,,	30	,,	16	,,	14	,,
,, 18	,,	26	,,	23	,,	3	,,
,, 19	,,	45	,,	29	,,	16	,,

Or adding them all together, while there were never less than ten pins, out of 156 ants, 103 came up the right pin, and only 53 up the others.

I was at first disposed to infer from these facts that the first ant must have described the route to its friends, but subsequent observations satisfied me that they might have found their way by scent.

APPENDIX E.

THE following are the details of the experiment referred to in p. 168 :—

January 24, 1875.—I put an ant, which already knew her way, on the larvæ at 3.22.
At 3.30 she returned.

4.15 ,,		At 3.38 another ant came; and the bridge *f* being turned towards *m*, she went over it to *m*.	
4.25 ,,			
4.34 ,,			
4.42 ,,	3.50	,,	,,
4.50 ,,	4.35	,,	,,
4.56 ,,	5.15	,,	,,
	At 5. 5 she returned.		
	5.14 ,,		
	5.25 ,,		

January 25.—6.30 A.M. put two ants, which knew their way, to the larvæ.

	No. 1.	No. 2.
Returned	6.55	
,,	7. 7	
		Returned 7.11
,,	7.15	
		,, 7.27
,,	7.35	
,,	7.46	
		,, 7.47

	No. 1.		No. 2.		
Returned	7.49				
		Returned	7.51		
,,	7.53				
		,,	7.57		
,,	8. 0				
,,	8. 3				
,,	8. 8				
				8.16 an ant to m.	
,,	8.17				
		,,	8.18		
,,	8.21				
				8.22	,,
,,	8.25	,,	8.25		
				8.27	,,
,,	8.29				
		,,	8.30		
,,	8.31				
		,,	8.34		
,,	8.35				
		,,	8.36		
,,	8.40	,,	8.40		
,,	8.44				
				8.45	
		,,	8.46		
,,	8.47				
,,	8.51	,,	8.51		
,,	8.55				
		,,	8 52		
,,	9. 3				
,,	9. 8				
,,	9.18				
,,	9.24				
,,	9.27				
,,	9.30				
,,	9.32				
,,	9.34				

	No. 1.		No. 2.	
		Returned	9.35	
Returned	9.37			
,,	9.43	,,	9.43	
				9.44 an ant to m
,,	9.45	,,	9.45	
,,	9.47			
,,	9.50			
		,,	9.51	
,,	9.55	,,	9.55	
,,	9 58	,,	9.58	
,,	10. 1	,,	1.10	
,,	10. 7	,,	10. 7	
,,	10.10	,,	10.10	
				10.11
		,,	10.15	
,,	10.16			
		,,	10.17	
,,	10.18			
,,	10.20	,,	10.20	
,,	10.22	,,	10.22	
,,	10.24			
,,	10.28			
		,,	10.30	
,,	10.32			
		,,	10.33	
,,	10.35	,,	10.35	
,,	10.38			
		,,	10.39	
,,	10.42	,,	10.42	
,,	10.45			
		,,	10.46	
,,	10.48			
		,,	10.49	
,,	10.51	,,	10.51	
,,	10.53	,,	10.53	
,,	10.55			

	No. 1.		No. 2.		
Returned	10.58	Returned	10.58		
,,	11. 0				
		,,	11. 1		
,,	11. 2				
,,	11. 5				
,,	11.10				
,,	11.12				
				11.15 another ant to m	
,,	11.16				
,,	11.21				
,,	11.23				
		,,	11.24		
,,	11.26	,,	11.26		
,,	11.30	,,	11.30		
,,	11.35	,,	11.35		
,,	11.36				
,,	11.40	,,	11.40	11.40	,,
				11.42	,,
		,,	11.43		
,,	11.45	,,	11.45		
,,	11.46				
		,,	11.50		
		,,	11.51		
		,,	11.56		
,,	11.58				
		,,	11.59		
,,	12. 0				
,,	12. 2	,,	12. 2		
,,	12. 6	,,	12. 6		
,,	12.10	,,	12.10		
,,	12.14				
,,	12.16				
,,	12.20	,,	12.20	12.20	,,
,,	12.24	,,	12.30	dropped.	
			1. 2	imprisoned her.	

POWER OF COMMUNICATION.

	No. 1.	No. 2.	
Returned	12.31	12.35	an ant **to** *m.*
,,	12.36		
,,	12.44		
,,	12.46		
,,	12.50		
,,	12.54		
,,	12.59		
,,	1. 1		

I then put her into a small bottle.

I let them out again at 7.10 on the 27th. Though the interval was so long, they began at once to work; but one unfortunately met with an accident. The other returned as follows, viz. at

7.20	
7.30	
7.40	
	7.48 stranger to *m.*
7.46	
7.51	
7.55	
7.59	

In these experiments, therefore, 17 unmarked ants came; but at the point *n* they all took the wrong turn, **and** **n**ot one reached the larvæ.

APPENDIX F.

THE following are the details referred to on p. 168:—
January 27, 1875.—At 5.30 I let out the same two ants as were under observation in the preceding experiments.

	No 1.			No. 2.	
Returned at	5.40,	the other not till	6.49		
,,	6. 0				
,,	6. 8				
,,	6.26			6.22 an ant to *m*.	
,,	6.32				
,,	6.37				
,,	6.41				
,,	6.45				
,,	6.48	,,	,,	6.49	6.50
,,	6.51				6.52 ,,
,,	6.54	,,	,,	7. 0	6.53 an ant to larvæ.
,,	7. 1				
,,	7. 5	,,	,,	7. 6	
,,	7. 9	,,	,,	7.12	
,,	7.17	,,	,,	7.17	
		,,	,,	7.22	7.27 an ant to *m*
,,	7.25	,,	,,	7.28	
,,	7.29	,,	,,	7.34	

I then put them into the bottle.

January 28.—Let them out at 6.45.

	No. 1.	No. 2.
Back at	7. 0	
,,	—	7. 3
,,	7. 5	
,,	7.11	
,,	—	7.12
,,	7.16	
,,	7.21	
,,	7.27	
		7.31 an ant to *m*.
,,	—	7.32
,,	—	7.42
,,	7.45	She dropped into
,,	7.52	some water.
,,	8. 2	
,,	8.11	
,,	8.20	
,,	8.26	
,,	8.30	
,,	8.36	
,,	8.40	
,,	8.44	
,,	8.48	

I then put them into the bottle.

January 29.—I let them out at 7.35 A.M.

No. 1 returned at 7.47, after which I saw her **no** more. I fear she must have met with an accident.

No. 2 returned at

7.56
8. 8
8.18
8.28
8.35
8.42
8.48

8.50 another ant came to the larvæ;
marked her No. 3.

8.56		
9. 5		
9.19	No. 3.	
	9.20	
9.26		
9.36		
9.46		2 ants to larvæ.
	9.47	5 ants to m.

At 9.40 I found one of the ants which had been under observation on the 24th, and put her to the larvæ. She returned as follows (No. 4):—

No. 2.	No. 3.	No. 4.
	9.50	
		9.52
	9.55	
9.58		
		10. 3
10.10		
		10.12
	10.15	
10.20		10.20
	10.23	
	10.26	10.26
	10.29	
	10.33	
	10.36	
10.37		
		10.40
10.41	10.41	
10.44		10.44
10.48		
		10.51
10.53		
	10.56	

No. 1.	No. 2.	No. 3.	
		10.57	
10.59	10.59		
	11. 2	11. 2	
11. 4			
			11. 5 an ant to larvæ.
	11.17		
			11. 8 ,, ,,
11. 9	11. 9		
		11.10	
	11.13		
		11.14	
	11.16		
11.17			
		11.18	
	11.20		11.20 ,, ,,
			11.21 ,, ,,
			11.22 an ant to m
11.23	11.23	11.23	
			11.25 an ant to larvæ.
	11.26		
11.28			
	11.30		
11.33	11.33		
			11.35 ,, ,,
11.40			
	11.42		
		11.44	
	11.46		
11.47			
	11.50	11.50	
	11.54	11.54	
			11.55 an ant to m.
	11.58	11.5	
12. 0			
	12. 1		
			12. 6 ,, ,,
		12. 7	

No. 1.	No. 2.	No. 3.
	12. 8	
12.10		
	12.13	
		12.14
12.15		
	12.18	
		12.24
	12.25	
12.27		
	12.30	
12.36	12.36	
		12.39
	12.40	
	12.43	
		12.45
	12.47	
	12.50	
		12.52
	12.53	
	12.56	
		12.57
	12.59	
		1. 0
	1. 7	1. 7
	1.12	
1.13		
		1.18
1.22		
		1.25
		1.53
		1.41
1.44		
		1.51
	1.55	
		1.56
		2. 9
		2.35

POWER OF COMMUNICATION.

I then put her into a small bottle. We kept a lookout for Nos. 2 and 3 till 7.30 p.m.; but they did not return.

January 30.—Let No. 4 out at 7 a.m. She returned at 7.45.

No. 3.	No. 4.
No. 3 came of herself at 8. 0	
Returning at 8. 9	8. 6
	8.15 another ant to larvæ.
,, 8.20	
	8.25
,, 8.30	
,, 8.36	
No. 3.	No. 4.
Returning at 8.40	
	8.43
	8.51 an ant to *m*.
,, 8.52	
	3. 3
,, 9 5	
Imprisoned them.	
Let them out at 10.55.	
Returning at 11. 1	
	11. 3
	11. 8
,, 11. 9	
	1'.14 another ant to *m*.

And they went on coming regularly till 1, when I put them again into a bottle.

January 31.—Let them out at 6.35 a.m.

No 3.	No. 4.
6.55	
7.12	
	7.15
7.21	

No. 3.	No. 4.	
	7.29	
7.37		
7.42	7.42	
7.48		
	7.53	
		7.55 another ant to *m*
		8. 0 ,,
8. 1		
8.12		
	8.18	
8.20		
		8.24 ,,
8.27		
	8.28	
8.32		
		8.36 an ant to larvæ.
8.39		
8.44		

I imprisoned them.

January 31.—Let them out at 5.35 P.M.

No. 3.	No. 4.	
	5.47	
6.25		
6.35		
6.48		
6.53		
7. 2		
7. 7		
7.11		
7.16		
7.20		
		7.23 another ant to larvæ
7.25		
		7.26 ,, ,,
		7.27 ,, *m.*

POWER OF COMMUNICATION. 395

No. 3.
                                7.29 another ant to *m*.
7.30   7.30   ,,   larvæ.
Imprisoned her   7.31   ,,   *m*..

February 1.—Let her out at 7.5.
             No. 3.
She returned at 7.20
    ,,   7.30
                          7.38 another ant to *m*.
    ,,   7.40
    ,,   7.48
    ,,   7.58
                      7.59   ,,
    ,,   8. 6
    ,,   8.12
                      8.14   ,,
                      8.17   ,,
    ,,   8.22

Imprisoned her and let her out again at 6.20 P.M.
She returned at 6.35
    ,,   6.52
    ,,   7. 0
    ,,   7. 5
    ,,   7.15
    ,,   7.20
    ,,   7.25
Imprisoned her.

February 2.—Let her out at 6.30 A.M.
She returned at 6.50
    ,,   7. 0
                      7. 2 another ant to *m*.
    ,,   7. 7
                      7.10 two other ants to *m*.
    ,,   7.13

She returned at 7.17

		7.27 another ant to larvæ.
,,	7.28	
,,	7.36	
		7.38 ,, m.
,,	7.45	
,,	7.50	
		7.51 ,, ,,
,,	7.55	
,,	8. 4	
		8. 6 ,, ,,
,,	8.11	
,,	8.18	
,,	8.25	
,,	8.30	
,,	8.35	
,,	8.45	
,,	8.46	

Imprisoned her.

In this experiment, then, the bridge over which the marked ant passed to the larvæ was left in its place, the scent, however, being removed or obscured by the friction of my finger; on the other hand, the bridge had retained the scent, but was so placed as to lead away from the larvæ; and it will be seen that, under these circumstances, out of 41 ants which found their way towards the larvæ as far as e, 14 only passed over the bridge f to the larvæ, while 27 went over the bridge d to the empty glass m.

Taking these observations as a whole, 150 ants came to the point e, of which 21 only went on to the larvæ, while 95 went away to the empty glass. These experiments, therefore, seem to show that when an ant has discovered a store of food and others flock to it, they are guided in some cases by sight, while in others they track one another by scent.

APPENDIX G.

The following are the details of the experiment referred to on p. 172:—

Experiment 1.—Time occupied, 1 hour. The ant with few larvæ made 6 visits and brought no friends. The one with many larvæ made 7, and brought 11 friends.

Experiment 2.—Time occupied, 2 hours. The ant with few larvæ made 13 journeys, and brought 8 friends. The one with many larvæ did not come back.

Experiment 3.—Time occupied, 3 hours. The ant with few larvæ made 24 journeys, and brought 5 friends. The one with many larvæ made 38 journeys, and brought 22 friends.

Experiment 4.—Time occupied, $2\frac{1}{2}$ hours. The ant with few larvæ did not come back. The one with many made 32 journeys, and brought 19 friends.

Experiment 5.—Time ocupied, 1 hour. The ant with few larvæ made 10 journeys, and brought 3 friends. The other made 5 journeys and brought 16 friends.

Experiment 6.—Time occupied, $1\frac{1}{2}$ hour. The ant with few larvæ made 15 journeys, but brought no friends. The other made 11 journeys and brought 21 friends.

Experiment 7.—I now reversed the glasses. Time occupied 3 hours. The ant with few larvæ made 23 journeys and brought 4 friends.

Experiment 8.—Time occupied, 1½ hour. The ant with few larvæ made 7 journeys and brought 3 friends. The one with many larvæ made 19 journeys and brought 6 friends.

Experiment 9.—Time occupied, 1 hour. The ant with few larvæ made 11 journeys and brought 1 friend. The one with many larvæ made 15 journeys and brought 13 friends.

Experiment 10.—I now reversed the glasses, the same two ants being under observation; so that the ant which in the previous observation had few larvæ, now consequently had many, and *vice versâ*. Time occupied 2 hours. The ant with few larvæ made 21 journeys and brought 1 friend. The one with many larvæ made 32 journeys and brought 20 friends. These two experiments are, I think, very striking.

Experiment 11.—Time occupied, 5 hours. The ant with few larvæ made 19 journeys and brought 1 friend. The one with many larvæ made 26 journeys and brought 10 friends.

Experiment 12.—Time occupied, 3 hours. The ant with few larvæ made 20 journeys and brought 4 friends. The one with many larvæ brought no friends and made 17 journeys.

Experiment 13.—Time occupied, 1 hour. The ant with few larvæ made 5 journeys and brought no friends. The one with many made 10 journeys and brought 16 friends.

Experiment 14.—I now reversed the glasses. Time occupied, 2½ hours. The ant with few larvæ made 10 journeys and brought 2 friends. The other made 41 journeys and brought 3 friends.

Experiment 15.—Time occupied, 4½ hours. The ant with few larvæ made 40 journeys and brought 10 friends. Of these, 8 came at the beginning of the experiment, and I much doubt whether they were brought; during the last hour and a half she only brought 1 friend. However, I think it fair to record the observation.

The ant with many larvæ made 47 journeys and brought 1 friend.

Experiment 16.—Time, 4½ hours. The ant with few larvæ made 20 journeys and brought 1 friend. She did not return after the first 2 hours. The other ant made 53 journeys and only brought 2 friends. This latter was the same one as in the previous experiment, when, however, she had the glass with only two or three larvæ.

Experiment 17.—Time, 1 hour. The ant with few larvæ made 6 journeys and brought no friend. The one with many larvæ made 11 journeys and brought 12 friends.

Experiment 18.—Time, 1½ hour. The ant with few larvæ made 25 journeys and brought four friends. The one with many larvæ made 20 journeys and brought 15 friends.

Experiment 19.—Time, 4½ hours. The ant with few larvæ made 74 journeys and brought no less than 27 friends. This is quite in opposition to the other observations; and I cannot account for it. She was the ant who brought 15 friends in the previous experiment, and it certainly looks as if some ants were more influential than others. The ant with many larvæ made 71 journeys and only brought 7 friends.

Experiment 20.—Time, 2 hours. The ant with few larvæ made 35 journeys and brought 4 friends. The one with many larvæ made 34 journeys and brought 3 friends.

Experiment 21.—I now transposed the two glasses. Time, 1½ hour. The ant with few larvæ made 15 journeys and brought no friends. The other made 35 journeys and brought 21 friends.

Experiment 22.—I now transposed the glasses again. Time, 2 hours. The ant with many larvæ made

37 journeys and brought 9 friends. The ant with few larvæ made 18 journeys and brought no friend. This, I think, is a very striking case. She was under observation 5½ hours; and the scene of her labour was the same throughout. The first 2 hours she had few larvæ and brought 4 friends; then for 1½ she had many larvæ and brought 21 friends; then again for 2 hours she had few larvæ and brought no friend.

Experiment 23.—Time, 1½ hour. The ant with few larvæ made 25 journeys and brought 3 friends. The other made only 9 journeys, but brought 10 friends.

Experiment 24.—I now transposed the glasses. Time occupied, 2 hours. The ant which now had few larvæ made 14 journeys, but brought no friends. The other made 37 journeys and brought 5 friends.

Experiment 25.—Time 3 hours. I put an ant for an hour to a full glass; she made 10 journeys and brought 4 friends. I then left only two or three larvæ: in the second hour she made 7 journeys and brought no friend. I then again filled the glass; and during the third hour she made 14 journeys and brought 3 friends

APPENDIX H.

The following are the detailed observations on bees alluded to in Chapter X.

August 24.— I opened the postern door at 6.45, and watched some marked bees till the middle of the day.

Bee No. 1.

6.50 One came to the honey. She then flew to the window, but after buzzing about for some time returned to the hive.
7.21 back to honey.   7.23 back to hive.
7.26 back to honey.
7.30 flew to window and then fell on the floor. I was afraid she would be trodden on, so at 7.45 I showed her the way to the hive.
8.40 back to honey.
8.45 back to hive. I now closed the postern door till 10.15.
10.35 back to honey.   10.39 to hive.
10.45   „   and then to hive.
12.35   „   12.37 to hive again.

Bee No. 2.

7. 0 she came to the honey. 7. 5 she went back to hive.
7.12 back to the honey.   7.22   „
7.24   „   7.30   „
7.42   „   7.46   „
7.52   „   7.57   „

8. 5 back to the honey.		8. 9 she went back to hive.	
8.15	,,	8.20	,,
8.26	,,	8.30	,,
8.40	,,	8.44	,,
8.55	,,	9. 0	,,

I then closed the door till 10.15; at 9.5, however, she came round to the honey through an open window, but could not find her way back, so I had to put her into the hive.

10.15 back to the honey.		10.17 she went back to hive.	
10.20	,,	10.23	,,
10.30	,,	10.33	,,
10.50	,,	10.55	,,
11. 1	,,	11. 6	,,
11.17	,,	11.23	,,
11.33	,,	?	,,
11.45	,,	11.50	,,
12. 0	,,	12. 3	,,
12.10	,,	12.15	,,
12.24	,,	12.30	,,
12.37	,,	12.43	,,
12.52	,,	12.56	,,

Bee No. 3.

Also on August 24.

10.16 came to honey.		10.19 returned to hive.	
10.30	,,	10.34	,,
10.55	,,	10.57	,,
11. 2	,,	11. 5	,,
11.11	,,	11.15	,,
11.24	,,	11.27	,,
11.35	,,	11.37	,,
11.45	,,	11.47	,,
11.57	,,	?	,,
12.13	,,	12.16	,,

COMMUNICATION AMONG BEES.

12.26 came to honey.   12.30 returned to hive.
12.36   ,,   12.42   ,,
12.56   ,,   12.59   ,,

The next day I timed this bee as follows:—

7.23 came to honey.   7.25 returned to hive.
7.35   ,,   7.37   ,,
7.44   ,,   7.45   ,,
8.10   ,,   8.12   ,,
8.53   ,,   8.55   ,,
                          (The door was then closed till 9.30.)
9.35   ,,   9.40 to window, and at
                                    9.49 to hive.
10. 0   ,,   10. 5 returned to hive.
10.13   ,,   10.15   ,,
10.22   ,,   10.26   ,,
10.35   ,,   10.40   ,,
10.45   ,,   10.48   ,,
10.56   ,,   ?   ,,
11. 7   ,,   11.12   ,,
11.18   ,,   11.20   ,,
11.35   ,,   11.37   ,,
11.47   ,,   11.51   ,,
12. 2   ,,   12. 6   ,,
12.25   ,,   12.29   ,,
12.51   ,,   12.54   ,,

During these observations scarcely any unmarked bees came to the honey.

In these cases the postern, being small and on one side, was not very easily found. If the honey had been in an open place, no doubt the sight of their companions feasting would have attracted other bees; but the honey was rather out of sight, being behind the hive entrance, and was, moreover, only accessible by the narrow and winding exit through the little postern door.

But, however exposed the honey might be, I found

similar results, unless the bees were visible to their fellows. Of this it may be well to give some detailed evidence.

Thus, one morning at

9.19 {	I brought a bee to some honey. }	9.24 {	she returned to the hive.
9.55 {	she came back to the honey. }	10. 0	,,
10. 8	,,	10.10	,,
10.16	,,	10.19	,,
10.28	,,	10.30	,,
10.37	,,	10.40	,,
10.50	,,	10.53	,,
11. 0	,,	11. 4	,,
11.11	,,	11.15	,,
11.22	,,	11.27	,,
11.34	,,	11.37	,,
11.46	,,	11.50	,,
11.55	,,	12. 0	,,
12. 6	,,	12. 7	,,
12.40	,,	12.46	,,
12.54	,,	12.57	,,
1. 2	,,	1. 4	,,
	Flew about.		
1.15	,,	1.18	,,
1.23	,,	1.27	,,
1.34	,,	1.41	,,
1.54	,,	2. 0	,,

After which she did not return. During this time no other bee came to the honey.

Again on another occasion I watched several bees, which on my list of marked bees stood as Nos. 3, 4, 7, 8, 10, and 11.

9.45 bee No. 10 came.		9.50 went back to hive.	
10. 0	,, 10 ,,	10. 3	,,
10.18	,, 10 ,,	10.21	,,
10.26	,, 11 ,,	10.30	,,

10.30 bee No. 4 came.		10.35 went back to hive.
10.36	„ 7 „	10.45	„
10.46	„ 4 „	10.52	„
10.49	„ 7 „	10.52	„
11. 0	„ 7 „	11. 9	„
11. 5	„ 4 „	11. 9	„
11.11	„ 7 „	11.16	„
11.21	„ 7 „	11.29	„
11.22 a strange bee came.			
11.26 bee No. 4 came.		11.31	„
11.30	„ 7 „	11.39	„
„	„ 10 „	11.36	„
11.40	„ 4 „	11.45	„
11.45	„ 7 „	11.50	„
11.47	„ 10 „	11.59	„
„ another strange bee came.			
12. 1 bee No. 4 came.		12. 6	„
12. 2	„ 7 „	12. 8	„
12. 3	„ 3 „	12. 7	„
12. 4	„ 10 „	12. 7	„
12.14	„ 7 „	12.18	„
12.17	„ 4 „	12.21	„
12.24	„ 7 „	12.31	„
12.30	„ 10 „	12.33	„
12.36	„ 7 „	12.46	„
12.37	„ 4 „	12.44	„
12.37	„ 10 „	12.40	„
12.45	„ 10 „	12.49	„
12.50	„ 7 „	12.54	„
12.50	„ 4 „	12.54	„
12.53	„ 10 „	12.56	„
12.57	„ 7 „	1. 0	„
12.57	„ 4 „	1. 2	„
1. 0	„ 10 „	?	„
1. 2	„ 7 „	1. 6	„
1. 9	„ 4 „	1.12	„
1.10	„ 8 „	1.16	„

1.10	bee No.	7	came	1.16	went back to hive.
1.16	„	4	„	1.19	„
1.17	„	5	„	1.21	„
1.20	„	7	„	1.24	„
1.20	„	8	„	1.25	„
1.21	„	4	„	1.24	„
1.23	„	5	„	1.27	„
1.29	„	4	„		
1.29	„	7	„		

After this I ceased recording in detail; but the above shows that while the marked bees came regularly, only in two cases did any unmarked bees come to the honey.

In the above cases the honey was poured into saucers, but not weighed. In the following I used a wide-mouthed jar containing rather more than one pound of honey.

1.44	bee No.	5	came.	1.45	went away.
1.54	„	5	„	1.58	„
2. 2	„	5	„	2. 5	„
2. 9	„	5	„	2.13	„
2. 9	„	1	„	2.15	„
2.18	„	5	„	2.20	„
2.19	„	1	„	2.21	„
2.28	„	1	„	2.31	„
2.37	„	1	„	2.41	„
2.32	„	5	„	2.40	„
2.49	„	5	„	2.51	„
2.52	„	1	„	2.55	„

3.10 another came which I numbered as No. 14

3.11	bee No.	1	came.	3.13	went away.
3.19	„	5	„	3.22	„
3.20	„	1	„	3.23	„
3.19	„	14	„	3.23	„
3.30	„	5	„	3.32	„
3.31	„	14	„	3.33	„

3.37 bee No.	1	came.	3.40	went away.	
3.38 ,,	5	,,	3.42	,,	
3.38 ,,	14	,,	3.41	,,	
3.47 ,,	5	,,	3.49	,,	
3.46 ,,	14	,,	3.51	,,	} She was
3.54 ,,	14	,,	3.56	,,	} disturbed.
4. 0 ,,	1	,,	4. 3	,,	
4. 0 ,,	5	,,	4. 3	,,	
4. 5 ,,	14	,,	4.11	,,	
4.10 ,,	5	,,	4.12	,,	
4.15 ,,	14	,,	4.20	,,	
4.22 ,,	1	,,	4.25	,,	
4.24 ,,	14	,,	4.29	,,	
4.26 ,,	5	,,	4.29	,,	

During the whole of this time only one strange bee came, as recorded above.

In the following case I put out, besides one pound of honey, also four ounces of honey spread over two plates.

12.15 one of my marked bees came.	12.21 she went.
12.26 she returned	12.31 ,,
12.36 ,,	12.44 ,,
12.51 ,,	12.57 ,,
1. 4 ,,	1.12 ,,
1.15 ,,	1.19 ,,
1.25 ,,	1.32 ,,
1.38 ,,	1.44 ,,
1.49 ,,	1.55 ,,
2. 0 ,,	2. 7 ,,
2.14 ,,	2.19 ,,
2.25 ,,	2.33 ,,
2.38 ,,	2.44 ,,
2.50 ,,	2.58 ,,
3. 5 ,,	3.13 ,,

3.20	she returned.	3.32 she went.	She was
3.39	,,	3.45 ,,	[disturbed.
3.52	,,	4. 0 ,,	
4. 7	,,	4. 9 ,,	
4.15	,,	4.20 ,,	
4.27	,,	4.32 ,,	
4.43	,,	4.45 ,,	
4.50	,,	4.59 ,,	
5. 7	,,	5.13 ,,	
5.25	,,	5.31 ,,	
5.42	,,	5.48 ,,	
5.56	,,	6. 1 ,,	
6.14			

During this time no other bee came to the honey.

I had, on August 20, introduced some bees to honey in my room, since which it had been much visited by them. On the 24th I put a bee to some honey inside a flower-pot five inches high and five wide at the base. The flower-pot was laid on its side, and the mouth closed, so that the bee had to come out through the hole in the bottom, which was about ½ an inch in diameter. To make things easier for her, I made her a small alighting-board of wood, the top of which was level with the hole. I then placed the flower-pot on the spot where she was accustomed to find the honey. She had made her first visit that morning at 6.45, returning

    At 6.55
       7. 5
       " 14
       7.23. I then arranged the flower-pot as described, and put her, while feeding, into it: she found her way out without difficulty.
    At 7.40 she returned, but did not seem able to find

her way; so I put her in. The same thing happened again at
7.50
8. 6
and 8.20
but at 8.38 she found her way in easily, and had **no** further difficulty. She returned at
8.53
9. 5
9.14
9.25
9.41
9.55
10. 6. This time a friend came with her and followed her in. I captured her. No. **2** took no notice, but returned

At 10.19	At 2.43
10.30	2.59
10.44	3.23
10.54	3.33
11. 6	3.44
11.20	3.56
11.31	4. 7
11.44	4.21
11.55	4.34
12. 9	4.44
12.25	4.55
12.37	5.10
12.50	5.24
1. 2	5.35
1.14	5.46
1.25	5.56
1.36	6. 9
1.47	6.20
1.57	6.42
2. 9	7. 0
2.19	7.15
2.31	making 59 visits.

After which she came no more that day. With the one exception above mentioned, during the whole time no other bee came to the honey. I might also mention that I had put out six similar flower-pots in a row, and that this seemed to puzzle the bee a good deal; she frequently buzzed about before them, and flew from one to the other before entering. When she went in, she generally stood still just inside the entrance for about thirty seconds, buzzing loudly with her wings. I thought at first whether this could be intended as a sort of gong to summon other bees to the feast; but though several were flying about, at any rate none came. The following day (August 25) she came at 6.51, and had made nine journeys up to 8.41, when I left off watching. During this time no other bee came.

August 26.—She came at 6.32, and up to 8.43 had made thirteen journeys.

August 27.—She came at 6.7, and up to 8.43 had made fourteen journeys.

August 28.—She came at 6.17, and up to 7.11 had made five journeys. During these days no other bee came.

On July 19 I put a bee (No. 10) to a honeycomb containing 12 lbs. of honey

At 12.30 she returned.		At 12.36 went back to hive	
12.50	,,	12.55	,,
1. 6	,,	1.12	,,
1.53	,,	1.57	,,
2. 5	,,	2 9	,,
2.16	,,	2.20	,,
2.28	,,	2.32 ?	,,
2.49	,,	2.55	,,
3.13	,,	3.20	,,
3.31	,,	3.39	,,
3.45	,,	3.58	,,
4. 2	,,	4. 8	,,

4.18	she returned.	4.24	went back to hive.
4.31	,,	4.37	,,
4.47	,,	4.58	,,
5.10	,,	5.19	,,
5.27	,,	5.30	,,
6. 9	,,	6.15	,,
6.23	,,	6.29	,,
7.19	,,	7.24	,,
7.35	,,	7.40	,,
7.50	,,	7.55	,,

and during all this time no other bee came to the comb.

On the following morning, July 20, this bee came to the honeycomb

At 6. 5 A.M.		At 6.10	went back to hive.
6.37	she returned.	6.42	,,
7.17	,,	7.21	,,
7.41	,,	7.47	,,
8. 8	,,	8.12	,,
8.21	,,	8.25	,,
8.32	,,	8.54	,,
9. 4	,,	9. 9	,,
9.45	,,	9.51	,,
10. 4	,,	10.10	,,
10.19	,,	10.26	,,
10.40	,,	10.47	,,
10.59	,,	11. 4	,,
11.14	,,	11.19	,,
11.44	,,	11.52	,,
11.59	,,	12. 6	,,
12.15	,,	12.23	,,
12.29	,,	12.35	,,
12.41	,, (was disturbed)	12.52	,,
1. 2	,,	1. 9	,,
1.16	,,	1.30	,,
1.46	,,	1.55	,,

I then left off observing; but during the whole of this time no other bee had come to the comb.

October 9.—I took a bee (No. 11) out of the hive and put her to some honey: she returned and kept on visiting it regularly.

October 10.—This bee came to the honey at 7.30 A.M., and went on visiting it; but I was not able to watch her continuously. During these two days no other bee came to this honey.

October 11.—No. 11 came to the honey

At 7.12 A.M., but did not alight.
7.18 she returned.   At 7.21 went back to hive
7.27   „   7.31   „
7.38   „   7.44   „
7.51   „   7.56   „
8. 2   „   8. 8   „
8.15   „   8.22   „
8.30   „   8.35   „
8.41   „   8.46   „
8.55   „   8.59   „
9. 6   „   9.11   „
9.20   „   9.25   „
9.45   „   9.50   „
9.55   „   10. 1   „
10. 7   „   10.11   „
10.19   „   10.23   „
10.30 a strange bee came; I killed her.
10.35 she returned. At 10.40 went back to **hive.**
10.55   „   10.59   „
11. 4   „   11. 8   „
11.26   „   11.30   „
11.35   „   11.38   „

Another strange bee came.

At 11.52 she returned. At 11.55 went.
12. 7   „   12.12   „

12.17 she returned.   12.22 went.
12.31   ,,   12.36 ,,
12.58   ,,   1. 2 ,,
1. 8   ,,   1.12 ,,
1.19   ,,   1.23 ,,
1.30   ,,   1.34 ,,
1.45   ,,   1.48 ,,
2. 2   ,,   2. 6 ,,
2.15   ,,   2.18 ,,
2.29   ,,   2.35 ,,
2.45   ,,   2.47 ,,
2.50   ,,   2.52 ,,
2.57   ,,   3. 0 ,,

after which she did not come any more that day. It was, however, a bad day, and after 1 o'clock she was almost the only bee which came out of the hive. The following morning she came to the honey at 7.58 A.M., but did not alight, behaving just as she had done the day before.

At 8. 6 A.M. No. 11 returned to honey. At 8. 9 went.
8.14   ,,   ,,   8.20 ,,
8.30   ,,   ,,   8.34 ,,
8.42   ,,   ,,   8.46 ,,
8.54   ,,   ,,   8.59 ,,
9. 9   ,,   ,,   9.14 ,,
9.19   ,,   ,,   9.24 ,,
9.29   ,,   ,,   9.33 ,,
9.37   ,,   ,,   9.44 ,,
9.54   ,,   ,,   but was disturbed.

A strange bee came. At 9.59 No. 11 went.
At 10. 5 she returned to the honey. At 10. 8 went.
10.12   ,,   ,,   10.13 ,,
10.16   ,,   ,,   10.20 ,,
10.26   ,,   ,,   10.28 ,,
10.33   ,,   ,,   10.36 ,,
10.40   ,,   ,,   10.46 ,,

10.55 a strange bee came. No. 11 returned to the honey regularly, and went on coming.

October 13.—At 6.28 A.M. she came, but, as before, flew away again without alighting.

At 6.32 she came to the honey. At 6.36 went away.

6.42	,,	,,	6.46	,,
6.51	,,	,,	6.56	,,
7.10	,,	,,	7.14	,,
7.26	,,	,,	7.34	,,
7.46	,,	,,	7.50	,,
7.55	,,	,,	8. 0	,,
8.12	,,	,,	8.15	,,
8.20	,,	,,	8.26	,,
8.30	,,	,,	8.33	,,
8.37	,,	,,	8.44	,,
8.50	,,	,,	8.56	,,

and so on.

October 14.—She came for the first time at 8.15 A.M., and went on visiting the honey at the usual intervals. After this day I saw her no more; she had probably met with some accident. But these facts show that some bees, at any rate, do not communicate with their sisters, even if they find an untenanted comb full of honey, which to them would be a perfect Eldorado. This is the more remarkable because these bees began to work in the morning before the rest, and continued to do so even in weather which drove all the others into the shelter of the hive. That the few strange bees which I have recorded should have found the honey is natural enough, because there were a good many bees about in the room. My room, I may add, is on the first floor; if it had been on the level of the ground I believe that many more bees would have found their way to the honey.

I will now proceed to the similar observations made with wasps.

The first one, I believe a worker of *Vespa Germanica*, I marked and put to some honey on September 18.

The next morning she came for the first time at 7.25, and fed till 7.28, when she began flying about the room, and even into the next; so I thought it well to put her out of the window, and she then flew straight away to her nest. My room, as already mentioned, had windows on two sides; and the nest was in the direction of a closed window, so that the wasp had to go out of her way in passing out through the open one.

At 7.45 she came back. I had moved the glass containing the honey about two yards; and though it stood conspicuously, the wasp seemed to have much difficulty in finding it. Again she flew to the window in the direction of her nest, and I had as before to show her the way out, which I did at 8.2.

At 8.15 she returned to the honey almost straight. At 8.21 she flew again to the closed window, and apparently could not find her way, so at 8.35 I put her out again. It seems obvious from this that wasps have a sense of direction, and do not find their way merely by sight.

At 8.50 back to honey, and 8.54 again to wrong window; but finding it closed, she took two or three turns round the room, and then flew out through the open window.

At 9.24 back to the honey, and 9.27 away, first, however, paying a visit to the wrong window, but without alighting.

At 9.36 back to honey; 9.39 away, but, as before, going first to wrong window. She was away, therefore, 9 minutes.

9.50 back to honey; 9.53 away.[1] Interval 11 minutes.
10. 0 ,,   ,, 10. 7 ,,   ,, 11 ,,

[1] This time straight.

10.19 back to honey; 10.22 away. Interval 12 minutes
10.35	,,	,,	10.39 ,,	,,	13 ,,
10.47	,,	,,	10.50 ,,	,,	9 ,,
11. 4	,,	,,	11. 7 ,,	,,	14 ,,
11.21	,,	,,	11.24 ,,	,,	14 ,,
11.34	,,	,,	11.37 ,,	,,	10 ,,
11.49	,,	,,	11.52 ,,	,,	11 ,,
12. 3	,,	,,	12. 5 ,,	,,	11 ,,
12.13	,,	,,	12.15½ ,,	,,	8 ,,
12.25	,,	,,	12.28 ,,	,,	10 ,,
12.39	,,	,,	12.43 ,,	,,	11 ,,
12.54	,,	,,	12.57 ,,	,,	11 ,,
1.15	,,	,,	1.19 ,,	,,	18 ,,
1.27	,,	,,	1.30 ,,	,,	8 ,,

Here for the first time another specimen came to the honey.

At 1.37 back to honey; 1.39 away.[1]
1.46	,,	,,	1.49 ,,	Interval	7 minutes.
1.54	,,	,,	1.58 ,,	,,	5 ,,
2. 5	,,	,,	2. 7 ,,	,,	7 ,,
2.15	,,	,,	2.19 ,,	,,	8 ,,
2.27[2]	,,	,	2.32 ,,	,,	8 ,,
2.39	,,	,,	2.42 ,,	,,	7 ,,
2.50	,,	,,	2.54 ,,	,,	8 ,,
3. 2	,,	,,	3. 6 ,,	,,	8 ,,
3.14	,,	,,	3.17 ,,	,,	8 ,,
3.26	,,	,,	3.29 ,,	,,	9 ,,
3.38	,,	,,	3.42 ,,	,,	9 ,,
3.50	,,	,,	3.58 ,,	,,	8 ,,
4. 7	,,	,,	4.12 ,,	,,	9 ,,
4.20	,,	,,	4.23 ,,	,,	8 ,,
4.32	,,	,,	4.36 ,,	,,	9 ,,

[1] Was rather disturbed, as I tried to mark her.
[2] She very often, however, throughout the day, in going away, flew to the other window first, and then, without alighting, returned to and went through the open one.

4.46 back to honey; 4.49 away. Interval 10 minutes.
5. 0 „   „   5. 3 „   „   11 „
5.13 „   „   5.17 „   „   10 „
5.26 „   „   5.30 „   „   9 „
5.40 „   „   5.44 „   „   10 „
5.54 „   „   5.59 „   „   10 „
6. 7 „   „   6.11 „   „   8 „
6.20 „   „   6.25 „   „   9 „

She did not come any more that day; but, as will be seen, she had made forty-five visits to the honey in eleven hours. During the whole of this time no strange wasp, except the one above mentioned, came to this honey.

The following day, September 20, this wasp made her appearance in my room at 6.55, when she flew straight to the honey.

At 6.55 came to honey; 6.59 went away.
 7. 8   „   7.10   „ Absent 9 minutes.
 7.18   „   7.22   „   „   8   „
 7.30   „   7.32   „   „   8   „
 7.41   „   7.45   „   „   9   „
 7.53   „   7.56   „   „   8   „
 8. 4   „   8. 7   „   „   8   „
 8.15   „   8.18   „   „   8   „
 8.27   „   8.30   „   „   9   „
 8.38   „   8.41   „   „   8   „
 8.50   „   8.53   ,   „   9   „
 9. 1   „   9. 4   „   „   8   „
 9.12   „   9.15   „   „   8   „
 9.22   „   9.25   „   „   7   „
 9.34   „   9.36   „   „   9   „
 9.46   „   9.51   „   „   10   „
10. 1   „   10. 3   „   „   10   „
10.13   „   10.18   „   „   10   „
10.28   „   10.30   „   „   10   „
10.38   „   10.42   „   „   8   „

10.53 came to honey; 10.56 away. Absent 11 minutes.
11. 7 　,,　　 11.11 　,,　 ,,　 11 　,,
11.21 　,,　　 11.25 　,,　 ,,　 10 　,,
11.32 　,,　　 11.36 　,,　 ,,　 7 　,,

The wasp which came once yesterday returned and rather disturbed the first.

At 11.49 came to honey; 11.50 away. Absent 13 minutes.
11.57 　,,　　 12. 0 　,,　 ,,　 7 　,,
12. 8 　,,　　 12.11 　,,　 ,,　 8 　,,

Here I was away for about two hours.

2.42 came to honey; 2.46 away.
2.58 　,,　　 3. 2 　,, Interval 12 minutes.
3.15 　,,　　 3.17 　,,　 ,,　 13 　,,
3.25 　,,　　 3.28 　,,　 ,,　 8 　,,

Here I was called away.

4.25 came to honey; 4.28 　,,
4.41 　,,　　 4.45 　,,　 ,,　 13 　,,
5.15 　,,　　 5.19 　,,　 ,,　 30 　,,
5.30 　,,　　 5.35 　,,　 ,,　 11 　,,
5.45 　,,　　 5.50 　,,　 ,,　 10 　,,
6. 2 　,,　　 6. 6 　,,　 ,,　 12 　,,
6.15 　,,　　 6.17 　,,　 ,,　 9 　,,

This was the last visit that day. She made, therefore, thirty-eight visits during the time she was watched, which was not quite eight hours. She was at work from 6.55 to 6.15; and assuming that she was occupied in the same manner during the three hours when she was not watched as during the rest of the time, she would have made over fifty visits to the honey during the day.

Wishing, however, to have a complete record of a day's work, I watched her the following day without intermission.

COMMUNICATION.

September 21.—I began watching at ten minutes past six.

6.16 came to honey; 6.19 away.

6.29	,,	6.32	,,	Interval 10 minutes.	
6.41	,,	6.44	,,	,,	9 ,,
6.55	,,	7. 0	,,	,,	11 ,,
7.11	,,	7.15	,,	,,	11 ,,
7.23	,,	7.26	,,	,,	8 ,,
7.37	,,	7.42	,,	,,	11 ,,
7.56	,,	8. 3	,,	,,	14 ,,

Was disturbed, and seemed rather troubled.

8.11 came to honey; 8.14 away. Interval 8 minutes.

8.20	,,	8.24	,,	,,	6 ,,
8.31	,,	8.34	,,	,,	7 ,,
8.40	,,	8.42	,,	,,	6 ,,
8.50	,,	8.52	,,	,,	8 ,,
8.58	,,	9. 0	,,	,,	6 ,,
9. 8	,,	9.11	,,	,,	8 ,,
9.18	,,	9.22	,,	,,	7 ,,
9.30	,,	9.32	,,	,,	8 ,,
9.39	,,	9.40	,,	,,	7 ,,
9.50	,,	9.54	,,	,,	10 ,,
10. 1	,,	10. 5	,,	,,	7 ,,
10.14	,,	10.17	,,	,,	9 ,,
10.25	,,	10.28	,,	,,	8 ,,
10.37	,,	10.40	,,	,,	9 ,,
10.47	,,	10.51	,,	,,	7 ,,
11. 0	,,	11. 6	,,	,,	9 ,,
11.17	,,	11.20	,,	,,	11 ,,
11.34	,,	11.37	,,	,,	14 ,,
11.50	,,	11.53	,,	,,	13 ,,
12. 5	,,	12. 8	,,	,,	12 ,,
12.20	,,	12.24	,,	,,	12 ,,
12.36	,,	12.40	,,	,,	12 ,,
1. 8	,,	1.11	,,	,,	28 ,,

1.26 came to honey;	1.28 away.		Interval	15	minutes
1.40 ,,	1.42 ,,		,,	12	,,
1.57 ,,	2. 2 ,,		,,	15	,,
2.10 ,,	2.13 ,,		,,	8	,,
2.25 ,,	2.30 ,,		,,	12	,,
2.45 ,,	2.56 ,,		,,	15	,,

She buzzed about at the other window for a few minutes, which made the interval longer than usual.

3.13 came to honey;	3.18 away.		Interval	17	minutes.
3.29 ,,	3.31 ,,		,,	11	,,
3.41 ,,	3.45 ,,		,,	10	,,
3.49 ,,	3.52 ,,		,,	4	,,
4. 2 ,,	4. 6 ,,		,,	7	,,
4.19 ,,	4.22 ,,		,,	13	,,
4.29 ,,	4.33 ,,		,,	7	,,
4.40 ,,	4.44 ,,		,,	7	,,
4.51 ,,	4.53 ,,		,,	7	,,
5. 4 ,,	5. 6 ,,		,,	11	,,
5.16 ,,	5.20 ,,		,,	10	,,
5.32 ,,	5.35 ,,		,,	12	,,
5.45 ,,	5.50 ,,		,,	10	,,

It will be seen that the intervals of her absence were remarkably regular. On one occasion, indeed, she was only away four minutes; but this time I think she had been disturbed, and had not provided herself with a regular supply of food.

The number of visits was fifty-one in eleven hours and a half. I tried whether she would be in any way affected by a dead wasp, so I put one on the honey; but she took no notice whatever.

I observed with other wasps, that when the open window was not the shortest way to their nests, they had a great tendency to fly to that which was in the right direction, and to remain buzzing about there.

INDUSTRY OF WASPS.

During the whole of this day only four or five strange wasps came to the honey.

As regards the regularity of their visits, and the time occupied, other wasps which I observed agreed very closely with this one. For comparison, it may be worth while to give one or two other cases. I will commence with that of a worker, I believe *V. vulgaris*, observed on September 19.

10 A.M. I put her to the honey; she fed and then flew about the room, and at last got into my beehive.

10.54. She came in again at the window. I again put her to the honey. She again flew all about the room.

11.41. She returned, and this time came to the honey; but when she had fed again flew round and round the room, and did not seem able to find her way out. I therefore put her out.

12.11 she returned, and the same thing happened again.

12.28 { came back to honey; } 12.31 flew straight away.

12.45	,,			
12.53	,,	12.57	,,	
1.10	,,			
1.26	,,	1.29	,,	
1.38	,,	1.41	,,	Interval 9 minutes
1.50	,,	1.53	,,	,, 9 ,,
2. 3	,,	2. 6	,,	,, 10 ,,
2.12	,,	2.16	,,	,, 6 ,,
		Was disturbed.		
2.20	,,	2.25	,,	,, 4 ,,
2.40	,,	2.43	,,	,, 15 ,,
2.51	,,	2.54	,,	,, 8 ,,
3. 1	,,	3. 4	,,	,, 7 ,,
3.13	,,	3.16	,,	,, 9 ,,
3.25	,,	3.28	.	9 ,,

3.35 { came back to honey; } **3.38** { flew away } Interval **7 minutes.**

3.46	,,	3.50	,,	,,	8 ,,
3.58	,,	4. 1	,,	,,	8 ,,
4.10	,,	4.14	,,	,,	9 ,,
4.23	,,	4.25	,,	,,	9 ,,
4.34	,,	4.38	,,	,,	9 ,,
4.46	,,	4.50	,,	,,	8 ,,
4.58	,,	5. 4	,,	,,	8 ,,
5.14	,,	Was disturbed and flew about.			8 ,,

She did not return any more that evening, but made her appearance again at half-past six the next morning.

From twelve o'clock, when she had learnt her way, till five, she made twenty-five visits in five hours, or about five an hour, as in the previous cases.

It struck me as curious that on the following day this wasp seemed by no means so sure of her way, but over and over again went to the closed window.

Again, September 21, at 11.50 I fed a wasp.

11.56	she returned to honey;		11.57	flew away
12. 6	,,	,,	12. 8	,,
1.25	,,	,,	1.27	,,
1.37	,,	,,	1.39	,,
1.57	,,	,,	2. 0	,,
2.15	,,	,,	2.17	,,
2.22	,,	,,	2.25	,,
2.32	,,	,,	2.36	,,
2.50	,,	,,	2.55	,,
3. 2	,,	,,	3. 4	,,
3.14	,,	,,	3.18	,,
3.28	,,	,,	3.30	,,
3.40	,,	,,	3.44	,,
3.51	,,	,,	3.55	,,
4. 4	,,	,,	4. 8	,,
4.16	,,	,,	4.20	,,

?	she returned to honey ;	4.31	flew away.
4.37	,,	,,	4.41	,,
4.46	,,	,,	4.48	,,
4.57	,,	,,	5. 0	,,
5. 9	,,	,,	5.12	,,
5.22	,,	,,	5.26	,,
5.31	,,	,,	5.36	,,

She made therefore twenty-three journeys, but did not bring a single friend.

The last case of which I will give particulars is the following, which has been already alluded to on p. 321. When I went to my sitting room at 4.13 A.M., I found her already there, though it was still almost dark. Her visits to the honey were as follows :—

4.13	A.M.,	returning	at
4.32	,,	,,	,,
4.50	,,	,,	,,
5. 5	,,	,,	,,
5.15	,,	,,	,,
5.22	,,	,,	,,
5.29	,,	,,	,,
5.36	,,	,,	,,
5.43	,,	,,	,,
5.50	,,	,,	,,
5.57	,,	,,	,,
6. 5	,,	,,	,,
6.14	,,	,,	,,
6.23	,,	,,	,,
6.30	,,	,,	,,
6.40	,,	,,	,,
6.48	,,	,,	,,
6.56	,,	,,	,,
7. 5	,,	,,	,,
7.12	,,	,,	,,
7.18	,,	,,	,,
7.25	,,	,,	,,
7.31	,,	,,	,,
7.40	,,	,,	,,

7.46 A.M., returning at
7.52 ,, ,, ,,
8. ,, ,, ,,
8.10 ,, ,, ,,
8.18 ,, ,, ,,
8.24 ,, ,, ,,
8.29 ,, ,, ,,
8.36 ,, ,, ,,
8.40 ,, ,, ,,
8.45 ,, ,, ,,
8.56 ,, ,, ,,
9. 7 ,, ,, ,,
9.14 ,, ,, ,,
9.20 ,, ,, ,,
9.26 ,, ,, ,,
9.37 ,, ,, ,,
9.43 ,, ,, ,,
9.50 ,, ,, ,,
9.57 ,, ,, ,,
10. 4 ,, ,, ,,
10.10 ,, ,, ,,
10.15 ,, ,, ,,
10.24 ,, ,, ,,
10.29 ,, ,, ,,
10.37 ,, ,, ,,
10.45 ,, ,, ,,
10.50 ,, ,, ,,
10.59 ,, ,, ,,
11. 6 ,, ,, ,,
11.15 ,, ,, ,,
11.22 ,, ,, ,,
11.30 ,, ,, ,,
11.35 ,, ,, ,,
11.47 ,, ,, ,,
11.55 ,, ,, ,,
12. 6 P.M., ,, ,,
12.14 ,, ,, ,,
12.22 ,, ,, ,,
12.36 ,, ,, ,,

12.46	p.m.,	returning	at
12.52	,,	,,	,,
12.56	,,	,,	,,
1. 4	,,	,,	,,
1.11	,,	,,	,,
1.20	,,	,,	,,
1.25	,,	,,	,,
1.30	,,	,,	,,
1.35	,,	,,	,,
1.43	,,	,,	,,
1.48	,,	,,	,,
1.53	,,	,,	,,
2.	,,	,,	,,
2. 7	,,	,,	,,
2.12	,,	,,	,,
2.23	,,	,,	,,
2.33	,,	,,	,,
2.39	,,	,,	,,
2.45	,,	,,	,,
2.55	,,	,,	,,
3. 2	,,	,,	,,
3. 9	,,	,,	,,
3.17	,,	,,	,,
3.25	,,	,,	,,
3.30	,,	,,	,,
3.37	,,	,,	,,
3.45	,,	,,	,,
3.55	,,	,,	,,
4. 5	,,	,,	,,
4.12	,,	,,	,,
4.19	,,	,,	,,
4.28	,,	,,	,,
4.39	,,	,,	,,
4.46	,,	,,	,,
4.56	,,	,,	,,
5. 3	,,	,,	,,
5.14	,,	,,	,,
5.25	,,	,,	,,
5.35	,,	,,	,,

5.46 P.M., returning at
5.50 ,, ,, ,,
6. 5 ,, ,, ,,
6.12 ,, ,, ,,
6.20 ,, ,, ,,
6.30 ,, ,, ,,
6.40 ,, ,, ,,
6.46 ,, ,, ,,
6.55 ,, ,, ,,
7. 7 ,, ,, ,,
7.17 ,, ,, ,,
7.30 ,, ,, ,,
7.36 ,, ,, ,,
7.46 ,, ,, ,,

This was her last visit for the evening, and she thus made no less than 116 visits in the day, during which time only three other wasps found the honey, though it was lying exposed on a table at an open window. It will be seen that she worked with the utmost industry.

No doubt, however, if a wasp is put to honey in an exposed place, other wasps gradually find their way to it. In the preceding experiments some few, though but few, did so. I then thought I would try a similar experiment with concealed honey. Accordingly, on September 20, I marked a wasp and put her to some honey, which she visited assiduously. The following morning I opened my window at 6, and she made her first visit at 6.27, the temperature being 61° Fahr. I then placed the honey in a box communicating with the outside by an india-rubber tube 6 inches long and ¼ inch in diameter. The wasp, however, soon got accustomed to it, and went in and out without much loss of time. The 22nd was finer; and when I opened my window at 6 in the morning, she was already waiting outside, the temperature being 61°. The 23rd was rather colder, and she came first at 6.20, the temperature being again 61°.

I was not at home during these days; but, as far as

A DAY'S WORK.

I could judge from watching in the mornings and evenings, no other wasp found the honey. On the 24th I had a holiday, and timed her as follows. It was rather colder than the preceding days, and she did not come till 6.40, when the temperature was 58°. She returned as follows:—

6.49	8.19
6.58	8.26
7.12	8.35
7.22	8.45
7.32	8.52
7.40	9. 2
7.50	9.12
8. 0	9.45
8. 9	

I had almost closed the window, so that she had a difficulty in finding her way.

9.58	10.32
10.10	10.51

The temperature was still only 60°, and it was raining, scarcely any other wasps about.

11. 1	1.42
11.11	1.53
11.21	2. 0
11.29	2.11
11.40	2.26
11.46	2.35
11.56	2.51
12. 6	2.59
12.14	3. 8
12.25	3.14
12.33	3.23
1.21	3.32
1.32	3.40

3.48	4.58
3.57	5. 6
4.12	5.17
4.20	5.28
4.29	5.35
4.39	5.42
4.47	5.52

This was her last visit. During the whole day no other wasp found the honey. I also tried other wasps, concealing the honey in the same manner, and with a similar result.

I have no doubt some wasps would make even more journeys in a day than those recorded above.

The following are descriptions of some new species referred to in the preceding pages. The first is the Australian honey ant.

CAMPONOTUS INFLATUS, n. sp.[1] *Operaria.*—Long. 15 mill. Nigra, tarsis pallidioribus; subtiliter coriacea, setis cinereo-testaceis sparcis; antennis tibiisque haud pilosis; tarsis infra hirsutis; mandibulis punctatis, hirsutis, sexdentatis; clypeo non carinato, antice integro; petioli squama modice incrassata, antice convexa, postice plana emarginata. *Hab.* Australian.

The colour is black, the feet being somewhat paler. The body is sparsely covered with stiff cinereo-testaceous hairs, especially on the lower and anterior part of the head, the mandibles, and the posterior edge of the thorax. The head and thorax are finely coriaceous.

The antennæ are of moderate length, twelve-jointed; the scape about one-third as long as the terminal portion, and somewhat bent. At the apex of the scape are a few short spines, bifurcated at the point. At the apex of each of the succeeding segments are a few much less conspicuous spines, which decrease in size

---

[1] In the *Linnæan Journ.* v. I have given figures of this species.

from the basal segments outwards. The antenna is also thickly clothed with short hairs, and especially towards the apex with leaf-shaped sense-hairs. The clypeus is rounded, with a slightly developed median lobe and a row of stiff hairs round the anterior border; it is not carinated.

The mandibles have six teeth, those on one side being rather more developed and more pointed than those on the other. They decrease pretty regularly from the outside inwards.

The maxillæ are formed on the usual type. The maxillary palpi are six-jointed, the third segment being but slightly longer than the second, fourth, or fifth; while in *Myrmecocystus* the third and fourth are greatly elongated. The segments of the palpi have on the inner side a number of curious curved blunt hairs besides the usual shorter ones.

The labial palpi are four-jointed. The eyes are elliptical and of moderate size. The ocelli are not developed.

The thorax is arched, broadest in front, without any marked incision between the meso- and meta-notum; the mesonotum itself is, when seen from above, very broadly oval, almost circular, rather broader in front and somewhat flattened behind. The legs are of moderate length, the hinder ones somewhat the longest. The scale or knot is heart-shaped, flat behind, slightly arched in front, and with a few stiff, slightly diverging hairs at the upper angles. The length is about two-thirds of an inch.

The following refers to a new species of mite which I have found in nests of *Lasius flavus*, and of which Mr. Michael has been good enough to draw up the following description.

UROPODA FORMICARIÆ, sp. nov.

This species, although it falls strictly within the genus *Uropoda*, and not within Kramer's genus *Trachynotus* as defined by that writer, still in most respects, except the very distinctions upon which the genus is founded, resembles *Trachynotus pyriformis* (Kramer) more closely than it does any other recorded species. It is, however, decidedly different, and is characterised by the squareness of its abdomen, the thickness and roughness of its chitinous dermal skeleton, and *especially* by the powerful chitinous ridges or wing-like expansions on the lateral surface between the second and third pair of legs.

Length, ♂ and ♀, about ·95 millim.
Breadth   "   "   ·55   "

The abdomen is almost square, but somewhat longer than broad, and slightly narrowed at its junction with the cephalothorax, from which it is not plainly distinguished. The extreme edge is a strong chitinous ridge bordered with a thick fringe of short, stout, curved hairs, as in *T. pyriformis*. The dorsal surface of the cephalothorax is also narrowed towards the front, and has a curved anterior margin bent down so as to protect the mouth, as in that species ; it bears a few of the same kind of hairs as the abdomen, and has a chitinous thickening at each side. The abdomen rises almost perpendicularly from the marginal ridge. There is a central depression occupying the posterior half, or rather more than half of the abdomen ; and at the bottom of this depression are transverse ridges, the hinder ones nearly straight, and the anterior ones bent in the middle, the central point being forward ; at the sides of, but not in, this depression, are two chitinous blocks which seem to form a starting-point for the ridges. Anterior to this depression the central portion of the creature, *i.e.* its longitudinal dorsal axis, is higher in level than in parts nearer the margin, and forms an irregular triangle of rough chitine. A broad chitinous plate or ridge projects on each side above the second leg, and between that and the third, evidently for their protection ; it is probably flexible at the will of the creature, as in the genus *Oribates*.

The sternal surface has strongly marked depressions for the reception of the legs. The coxæ of the first pair of legs are largely developed, flattened, almost touch in the median line, and nearly conceal the mouth, as in the typical *Uropodus*. The genital opening of the male is rather large, round, and placed centrally between the coxæ of the second pair of legs. The female appears only to be distinguished from the male by being more strongly chitinised, and by the conspicuous valval plate which occupies the whole space between the coxæ of the second and third pairs of legs and extends beyond both.

The nymph is less square in the abdomen than the adult, and the border of hairs is absent; the margin is somewhat undulated, the concave undulations being so placed as to give free action to the legs when raised; the central depression of the abdomen is far less marked than in the adult; a slight ridge runs all round the dorsal surface a little within the margin; four ridges, two anterior and two posterior, run from the circumscribing ridge to a raised ellipse in the centre; there are not any plates for the protection of the legs, and the coxæ of the first pair are not flattened as in the adult.

This mite lives in the nests of *Formica flava*.

Description of a New Genus and Species of Phoridæ parasitic on Ants. By G. H. VERRALL, Esq., Memb. Entom. Soc.

Sir JOHN LUBBOCK has kindly forwarded for my examination and determination certain specimens of dipterous insects said to have been found parasitic on species of ants, which latter he has been studying with care as to their habits. Having given considerable attention to the family Phoridæ, I was agreeably surprised to find the parasitic specimens to be forms new to science. One of these is a new species of the genus *Phora;* the other I regard as possessing characters *sui generis*, and hence define it under the generic title *Platyphora*, at the same time bestowing on the species the name of the discoverer, who worthily pursues entomological researches, spite of many pressing public engagements.

The subjoined descriptions embrace the diagnostic peculiarities of the insects in question.

PHORA FORMICARUM, n. sp.—Nigro-cinerea, fronte setosa, caniculata; antennis mediocribus, cinereis; palpis magnis, flavis; halteribus flavidis; pedibus totis pallide flavis, inermibus, tibiis intermediis unicalcaratis, posticis modice dilatatis; alis subhyalinis, nervo secundo simplici, nervulis vix undulatis. Long. vix ½ lin.

Frons broad, grey, bristly, two large bristles being close to the eye-margin; down the centre is a deep impressed channel, which at its lower end joins a channel above the antennæ, and at its upper end a channel round the raised vertical triangle; the space between these two latter channels (comprising the true *frons*) is about once and a half broader than deep; on the vertical triangle are two bristles; the third joint of the antennæ is moderately large, ovate, grey; the arista short, somewhat yellowish, almost naked; the palpi conspicuous, all pale yellow, with a few short black bristles at the tip; on the cheeks are some short black bristles.

The thorax is grey or brownish grey, broad, not much arched, the disk being nearly flat, and on the hinder part absolutely concave; on the disk there are no long bristles, but a dense clothing of rather short black bristles; along the side of the thorax between the humeri, the base of the wing, and the scutellum are some long black bristles, and two on the thorax just before the scutellum; on each side of the scutellum are two long bristles; halteres dirty pale yellow; abdomen bare, dull black, with slightly yellowish incisures; ovipositor polished black, long, slightly incurved and grooved.

Legs pale yellow, including the coxæ, clothed with minute black bristles; all the coxæ with two or three black bristles at the tips, the legs otherwise bare excepting the spurs; femora flattened and widened, especially the hind pair, the hind tibiæ also slightly flattened and widened on the apical half; middle tibiæ with a long spur inside at the tip, and hind tibiæ with a small one inside and a very minute one outside; tarsi longer than the tibiæ, joints gradually diminishing in length.

Wings very slightly smoky, broad; second thick vein not extending half the length of the wing, thickened, but not forked at its tip; first veinlet with a steady curve; second very slightly curved at base, otherwise straight; third very slightly undulated; fourth hardly visible at base, evident towards tip, very slightly undulated; costa bristly up to end of second thick vein.

This species is readily distinguished by its simple second thick vein, channelled frons, small size, and by the absence of bristles on the tibiæ.

It is parasitic on *Lasius niger*.

PLATYPHORA, n. gen.

Lata, planx, tota absque setis. Frons latissima. Thorax transversus. Abdomen parvus. Alarum vena cubitalis simplex, subacostali parallela; venulæ undulatæ; costa ad basin subciliata.

Distinguished from all the existing genera of Phoridæ by its flat and broad shape, which resembles that of the small species of *Sphærocera*. The absence of strong bristles on the frons, thorax, and legs also distinguishes it from all the genera except *Gymnophora*, which, however, is of the usual arched *Pho-*

ra-shape, and has the cubital vein forked, costa bare, &c.

PLATYPHORA LUBBOCKII.—Nigra, nitida; abdomine triangulari, segmento tertio parvo; femoribus posticis basi flavidis; alis apice latis, flavido-hyalinis, costa ad basin subciliata, vena cubitali ad medium costæ extensa subcostali parallela, venulis undulatis. Long. ¾ lin.

Broad, flat, shining; frons very broad, the eyes scarcely occupying each one-sixth the width of the head; it is moderately shining, gently arched, and pretty densely clothed with minute bristles; the three ocelli visible slightly luteous; antennæ with the third joint rather large, somewhat rounded; thorax broad, flat, rather broader than the head, angles tolerably rounded, disk shining (in appearance suggesting a small *Sphærocera*), beset with very minute bristles, which become rather scarcer towards the hinder part; scutellum rather dull, margined, nearly four times as broad as long: abdomen black, narrower and shorter than the thorax (again suggestive of *Sphærocera*); each segment after the second successively narrower, the last one being almost triangular; the third segment is very short, contracted under the second; the hind margins form a curved convex towards the thorax, the first segment being slightly emarginate in the middle; the sixth (last) is much the longest. Legs stoutish, blackish, basal two-thirds of hind femora yellowish; middle tibiæ with two small spines at the tip. Wings considerably overlapping the abdomen, yellowish hyaline, darker about the basal half of the costa, blunt at the tip, cubital vein extending about half the length of the wing, and the costa slightly ciliate up to its end, subcostal vein running parallel to it and ending just before it; both veins a little thickened at their ends; first veinlet curved S-like, considerably at its base, slightly at its end, vanishing distinctly before the tip of the wing; second veinlet also S-like, diverging at its end from the first, and ending distinctly below the tip of the wing; third veinlet slightly undulated, ending very wide from the second; fourth faint, not reaching the end of the wing.

This description having been made from a specimen gummed down on card, though in very good condition, I am unable to decide on the sex, or to examine the face, palpi, base of antennae, or coxae.

INDEX.

ABD

ABDOMEN of ant described, 10, 13; of the Mexican honey ant, 19, 47
Acacia with hollow thorns inhabited by ants, 57
Affection less powerful than hatred among ants, 106; absence of, among bees, 286
Agricultural ants, 61, 92
Aldrovandus quoted as to ants, 61
Amazon ants, see *Polyergus rufescens*
Amber, an intermediate form of ant preserved in, 68
Analogies between ant societies and human, 91
André quoted as to *Platyarthrus*, 75; as to the slaves of *F. sanguinea*, 80
Anergates, 85; no workers among them, 86; degraded condition of, 89
Animal food, queens hatched in an artificial nest supplied with, 40
Angræcum sesquipedale, length of flowers of, 52
Anomma arcens, the Driver ant, described, 20, 63; their blindness, 65
Ants, three families of, 1; four periods of life in, 6; duration of life among, 8, 38, 40; structure

AFH

of, 10; different classes of individuals among, 18; communities of, 24; games of, 28; their relation to plants, 50; often insectivorous, 59; their relations to other animals generally hostile, 63; their enemies, 26, 67; their domestic animals, 67-78; progress among, 90; their behaviour towards each other, 94, &c.; mental powers of, 181; their sense of vision, 11, 182-220, 258; of smell, 127, 238, 258; of hearing, 221, 226; stridulating apparatus among, 230, their intelligence, 236
'Ant eggs,' 7
'Ant-rice,' 61
Antenna of ant described, 10; sense organ in terminal portion of, illustrated. 227
Antennæ as means of communication among ants, 153; as organs of hearing, 221, 226; of smell, 94, 234
Antirrhinum fertilised by humble bees, 54
Aphides made use of by ants, 25, 67; different species of, utilised by different ants, 68; their honey, 69; their eggs tended by ants, 70; not domesticated by *F. fusca*, 91

ARI

Aristida oligantha, 'ant-rice,' 61
Artificial nests for ants, 3, 164
Ateuchus pilularius, anecdote of, 154
Atrophy of the imaginal discs of the ant-workers, 12; of the sting in *Formica*, 15; of the eyes of *Platyarthrus* and *Beckia*, 75
Atta barbara, the eye in, 11; variety of workers among, 19
— *structor*, its treatment of collected grain, 61
— *testaceo-pilosa*, experiment with, as to power of communication, 177
Attachment among ants, 94
Auditory organs, structures in ant-antennæ probably serve as, 226
Australian honey ant, 49; described, 428

BATES, Mr., quoted as to the five kinds of workers in Saüba, 22; as to ant-play, 29; as to the use made by ants of leaves, 57; as to the armies of *Eciton*, 65; as to leaf-cutting by Saüba, 237
Batrisus, rarely more than one specimen of, found in an ants' nest, 78
Beckia, one of the ant-guests, 74
Bees, occasional fertility of workers among, 36; means of recognition among, 126; their sense of hearing, 221, 290; observations with, 274; difficulty in finding their way, 278; their behaviour in a strange hive, 281; their recklessness, 285; their want of mutual affection, 286; their influence on the development of flowers, 51, 292; their colour sense, 291; their preference for blue, 294, 310; experiments on communication among them, 276, 401

CAR

Beetles kept in ants' nests, 74, 76, 90
Belt, Mr. Thomas, quoted as to floral defences against ants, 51; as to defence against leaf-cutting ants, 57; on the raids of *Eciton*, 66; on an ant-like spider, 66
Bert, Prof. Paul, as to the limits of vision, 219
Bichromate of potash, experiments with, 211
Bisulphide of carbon, experiments with, 208;
Blanchard, M., quoted as to the origin of nests, 30
Blindness of *Anomma* and *Eciton* 65; of *Platyarthrus* and *Beckia*, 75
Blue, the favourite colour of bees, 294, 304, 310; flowers, their late origin, 308
Bonnet, M., on aphis eggs, 70
Bonnier, M., on indifference to colour among bees, 302
Bothriomyrmex meridionalis, the eye in, 11
Brazil, blind hunting ants of, 65; use made by the Indians in, of the tenacity of an ant-bite, 96
Buchlæ dactyloides, seed of, collected by ants, 61
Büchner, Dr., as to Texan harvesting ants, 62
Burmeister, on the power of recognition among insects, 126
Butterfly, ants seen licking the larva of, 68

CAMPONOTUS *inflatus*, described, 428
— *ligniperdus*, the eye in, 11; communication among 158
Captivity, mode of keeping ants in, 2, 3; a wasp in, 315
Caterpillars killed by ants, 59, 65
Caryophyllaceæ, correlation of form and colour in, 309

CHE

Chennium, rarely more than one specimen of, in an ants' nest, 78
Christ, M., on the length of life of queen ants, 9; on ant roads, 25
Chrome alum. experiments as to ant vision with, 217
Chromium chloride, experiments with, 217
Claparède, M, as to insect-vision, 183
Clark, Rev. Hamlet, as to an anttunnel in S America, 25
Claviger, a blind beetle, a guest in ants' nests, 75, 76; experiments with, by M. Lespès, 90
Cleanliness of ants, 29
Coccidæ, their use to ants, 68
Cocoons spun by some larvæ of ants, 7
Colobopsis truncata and *C. fusipes*, two forms of the same species, 20
Colour-sense of ants, 186, &c.; of bees, 291, &c.; of wasps, 316; less developed among wasps than bees, 321
Colours of flowers, evolution of, 308
Communication, power of, among ants, 153, &c.; among bees, 156; experiments as to, with ants, 160, 344, 376; with bees, 276, 401; with wasps. 311, 415
Communities of ants, 24; power of mutual reco_nition among members of, 119, 333
Compassion among ants, instances of, 106, 108; absence of, among bees, 286
Co-operation, experiments as to, among ants, 365-376
Correlation of form of knot with stinging power in ants, 13; of colour in flowers with specialisation of form, 308
Courage of ants, 27; of wasps, 314
Crematogaster lineolata, adoption of a queen by, 34

DUJ

Crematogaster scutellaris, their neglect of friends in trouble, 98; experiments as to perception of colour among, 192
— *sordidula*, threatening attitude of, 16
Cross-fertilisation effected by insects, 50

D*APHNIA*, limits of vision in, 219
Darkness, education of young ants conducted in, 5; effect of, on the eyes of *Platyarthrus* and *Beckia*, 75
Darwin, on the sound produced by *Mutilla*, 229
— Francis, on the use of the leafcups of teazle, 52
Dead, treatment of the, among bees, 287
Defences of flowers against unbidden guests, 52-7
Degradation of *Strongylognathus*, 85; caused by slaveholding, 89
Dewitz, Dr., on the non-development of the sting in the Formicidæ, 14; on eggs laid by fertile workers, 36, 40
Dinarda dentata in ants' nests, 76, 77
Dipsacus sylvestris, leaf-cups of, 52
Direction, sense of, among ants, 260; guided by the position of the light, 268; sense of, among bees, 278; among wasps, 321, 420
Discs, atrophy of imaginal, in worker ants, 12; cleared by harvesting ants, 61: experiments as to sense of direction with rotating. 261, &c.
Division of labour among ants, 23, 44; tabular view of experiments on, 324
Domestic animals of ants, 68-78
Driver ants, see *Anomma arcens*
Dujardin, M., as to the power of

EBR

communication among bees, 156. 313

EBRARD, M., his observations as to the origin of ants' nests, 31
Eciton, the eye in, 11
— *drepanophora*, their order in marching, 21
— *erratica*, soldiers among, 21; their covered galleries, 65
— *legionis* at play, 29
— *rastator*, soldiers among, 21; their covered galleries, 65
Economy of labour among ants, experiments as to, 240, &c.
Eggs of ants described, 6; laid occasionally by worker ants, 35; by worker bees and wasps, 36; these always produce males, 37; as to difference of sex in, 40; of aphis, tended by ants, 69; and hatched in captivity, 71
Electric light, experiments on ants with, 200
Emery's observations on *Colobopsis*, 20
Enemies of ants, 26, 67
Evolution of colour in flowers, 308
Experiments, as to the adoption of a queen by ants, 32; as to division of labour among ants, 23, 44, 324; as to their care of aphis-eggs, 70; on *Claviger*, 90; as to the treatment by ants of injured companions, 94, 107; with chloroformed ants, 98, 108-111; with drowned ants, 99; with buried ants, 102; as to treatment of stranger ants, 104, 119, 124, 333; as to mode of recognition, 108; with intoxicated ants, 111 118; as to power of recognition among ants, 119, 333; and among bees, 126; with ant-pupæ removed from nest, 129-147; on sister-ants brought up separately, 117-

FOO

152, as to power of communication among ants, 160-181, 344-376; among bees, 274, 401; among wasps, 311, 415; as to perception of colour, 186; with coloured solutions, 194; with spectrum, 198; with the electric light, 201; as to ultra-violet rays, 200-220; with magnesium spark, 207; as to sense of hearing among ants, 222; among bees, 290; as to sense of smell among ants, 233, 258; among bees, 288; as to ant intelligence, 237; as to economy of labour, 241; as to ingenuity among ants, 243-6; as to their power of finding their way, 250; as to means of tracking, 168, 383, 387; as to sense of direction among ants, 260; among bees, 278; and among wasps, 321; as to guidance of ants by sight, 266; as to the behaviour of bees in a strange hive, 281; as to their compassion, 286; as to their colour sense, 291; and their preference of certain colours, 302; as to colour sense among wasps, 316
Expulsion of ant from nest, 98
Eyes of two kinds in ants, 10; compound, 182; various developments of, 183

FACETS of the eye in ants, number of, 11; described, 182
Feeding, loss of instinct of, 76, 83, 87
Fertilisation of plants by insects, 50, 291
Fighting among ants, different modes of, 17
Flowers, their defences against unprofitable insects, 51-55; influence of bees on their development, 291; paucity of blue, 308
Food of ants, 25, 65; its effect in

INDEX. 441

FOR

determining the sex in ants and bees, 40; individual ants in certain species serve as receptacles of, 47

Foragers, certain ants of a nest told off as, 45, 47

Forel, Dr., referred to as to the emergence of pupæ of ants, 8; as to their compound eyes, 10; as to the position of spiracles, 14; as to the offices of young ants, 23; as to *F. rufa*, 27; as to ant-games, 28, 29; as to origin of nests, 31; as to eggs laid by workers, 35; on the honey ant, 40; on the germination of grain in ant-stores, 61; as to beetles in ant nests, 78; as to the slaves of *F. sanguinea*, 80; as to the slave-making of *Strongylognathus*, 85; on *Anergates*, 86; on the behaviour of ants to each other, 94; on recognition among ants, 120; as to power of communication among ants, 158; as to their insensibibility to sound, 221; as to special organs in their antennæ, 227

Formica bispinosa, its nest, 24
— *cinerea*, 16; character of, 27; eggs laid by workers among, 37, 39; duration of life of, 42
— *congerens*, *Thiasophila* in nests of, 77
— *exsecta*, mode of attack of, 17; extent of nest of, 24; *Thiasophila* in nests of, 77
— *flava*, *Uropoda* in nests of, 431
— *fusca*, occasionally spins a cocoon, 7; its timidity, 27; introduction of a queen among, 34; eggs laid by workers among, 38, 39; queens produced in captivity, 40; longevity of, 42; division of labour among, 45; occasionally found in the nests of *F. rufa*, 79; enslaved by *F. sanguinea*, 80; *Platy-*

FOR

arthrus received in nests of, 90; their condition analogous to that of the hunting races of men, 91; their neglect of friends in trouble, 96; expulsion of a member from the nest, 98; mite attached to the head of a queen of, 98; their neglect of imprisoned companions, 103; hostility towards imprisoned strangers, 104; instances of their kindness to crippled companions, 106; experiments as to recognition among, 122, 130, 134, 233; on power of communication among, 161, 180; as to perception of colour among, 188, 193, 201

Formica gagates enslaved by *F. sanguinea*, 80
— *ligniperda*, experiments as to sense of hearing among, 223; as to sense of smell among, 234
— *nigra*, experiment as to power of communication among, 363
— *pratensis*, eye of, 10, 184; attacked by *F. exsecta*, 18; its treatment of slain enemies, 27; *Stenamma* in nests of, 78; large communities of, 119
— *rufa*, its power of ejecting poison, 15; its mode of attack, 17, 27; nests of, 23; large number of insects kept in nests of, 74, 75; *Stenamma* in nests of, 78
— *rufibarbis* perhaps a variety of *F. fusca*, 80
— *sanguinea*, its mode of attack, 17; duration of life of, 41, 42; *Dinarda* in nests of, 77; their periodical attack on neighbouring nests, 79; slaves made by, 80; not yet degraded by slave-holding, 88; they apparently understand the signals of *Pratensis*, 159

Formicidæ, one of the three families of ants, 1; power of stinging absent in them, 13

FRA

Franklin, Dr., as to power of communication among ants, 155
Friends, behaviour of ants to, 97, 101; recognition of, 119, 333

GALLERIES, covered, made by *Eciton*, 65
Galton, Mr. Francis, on domestic animals kept as pets, 77
Games among ants, 28
Gélieu, M., on means of recognition among bees, 126
Gentians, colours of, 310
Glasses, experiments on ants with coloured, 186, &c.
Gould, Mr., on the emergence of the imago, 8; on ant-games, 28; on the eggs of aphides, 69
Goureau, M., on the sound produced by *Mutilla*, 229
Graber, Dr. von, on the sense organs in the legs of *Gryllus*, 231
Grain collected and stored by ants, 26, 60; germination of, prevented by ants, 61
Gredler, Dr., anecdote of ant-intelligence told by, 237
Grimm's observations on *Dinarda*, 76
Grote, Mr., quoted as to the necessity of morality in societies, 93
Growth of insects takes place during the larval stage, 8
Guests of ants, 74
Gryllus, sense organs in tibiæ of, 231

HABITATIONS of ants, 24
Hagens, von, quoted as to myrmecophilous beetles, 77; on the slave-making of *Strongylognathus*, 85; on *Anergates*, 86; his suggestions as to *Strongylognathus*, 87
Hairs of p ants as defences against insects, 55
Harvesting ants, 59, 92; in the South of Europe and Texas, 61

INS

Hatred a stronger passion with ants than affection, 106
Head of ant described, 10; large size of, in workers of certain species, 20, 22
Hearing, sense of, among insects, 221; possibly present in ants, 226; among bees, 290; among wasps, 313
Hetærius sesquicornis, in ants' nests, 77
Hicks, Dr. J. Braxton, on the antennæ of insects, 227
Hildebrand on the variations of blue flowers, 310
Honey, love of ants for, 51; of aphis, 69; experiments on ants entangled in, 98
Honey ants, 19, 47; independently originated in Mexico and Texas, 49; of Australia described, 428
Hope, Mr., quoted as to harvesting ants, 60
Horse ant, see *F. rufa*
Huber, as to ants playing, 28; as to the formation of a nest, 30; on the care taken by ants of aphis eggs, 70; as to slavery among ants, 81; as to their recognition of friends, 120; as to their deafness, 221
Hunting ants, 59, 63, 91
Hydnophytum formicarum, its association with ants, 58
Hymenoptera, common origin of the sting in the, 15; the social, means of communication between, 153

INDIVIDUAL differences between ants, 95, 101; between bees, 279
Industry of ants, 27; of wasps, 321, 421
Insects, their metamorphoses, 8; their agency in fertilisation of flowers, 50, 291; mimicking ants, 66; kept by ants, 73; recognition among, 126; their vision,

182; their hearing, 221; possibly possess senses inconceivable to ourselves, 225
Insensible ants, experiments with, 99–108
Intelligence among ants, 181, 236; experiments as to, 240
Intoxicated ants, experiments on, 111; tabular view of experiments, 118; experiments referred to, 128

KERNER on floral defences, 52; on the uses of nectaries, 56
Knot in ants, specific characters offered by form of, 13
Kirby and Spence, Messrs., as to power of communication among ants, 156; on the power of sound in *Mutilla*, 229

LABOUR, division of, among ants, 23, 44; experiments as to economising, 240; tabular view of experiments on, 324–332
Landois, on the sound emitted by *Mutilla*, 229; on stridulating apparatus in ants, 230
Langstroth, Dr., as to recognition by smell among bees, 281; on their recklessness, 285
Larvæ of ants described, 6; of stranger nests carefully tended, 129
Lasius brunneus prefers the aphides of the bark of trees, 68
— *flavus*, period of larval life in, 7; the eye in, 11; will not adopt a strange queen, 32; mites in the nest of, 67; keeps flocks of the root-feeding aphis, 68; keeps four or five species of aphis in its nests, 73; *Platyarthrus* a guest of, 75, 90; they have arrived at the 'pastoral stage' of progress, 91; their behaviour to a dead queen, 108; to chloroformed friends and strangers, 108–111; to intoxi-

cated friends and strangers, 111; their treatment of strangers, 123; perception of colour among, 190, 193, 195; experiments as to sense of hearing among, 223; stridulating apparatus in, 231; structure in tibia of, 232; want of ingenuity among, 248; earthworks constructed by, 249; experiments with, as to power of communication, 365; as to cooperation, 372; new species of mite found in nests of, 429
Lasius niger, workers among, 19; fed by aphides, 25; eggs laid by workers among, 37; longevity of, 42; typical nest of, illustrated, 42; and described, 44; they carry seeds of violet into their nests, 59; their choice among aphides, 68; *Platyarthrus* a guest of, 75; *Hetærius* found in nests of, 77; experiments as to *Claviger* in nests of, 90; observations on a wounded worker among, 95; experiments with buried individuals of, 102; with pupæ as to recognition among, 131; as to power of communication among, 160, 163, 172, 175, 356–362, 377; as to perception of colour among, 191, 201; as to their intelligence, 240; their want of ingenuity, 242, 246; as to scent, 258; as to sense of direction among, 260; *Phora formicarum* parasitic on, 433
— *fuliginosus*, the eye in, 11; stridulating apparatus in, 230
Latreille quoted as to compassion shown by ants, 94
Leaf-cutting ants, 57
Legs of ants described, 12
Leptothorax acervorum, *Platyarthrus* a guest of, 75; *Tomognathus* in nests of, 87 *note*
— *muscorum*, *Tomognathus* in nests of, 87 *note*

LES

Lespès, M., on eggs laid by workers, 35; on grain stored by ants, 61; on the feeding of *Lomechusa* by ants, 76; on the domestic arimals of ants, 90
Leuckart, his experiments on ant-intelligence, 238
Life, duration of, among ants, 8
Light, dislike of ants to, 2, 186; as aid to sense of direction among ants, 268; bees attracted by, 284
Ligurian queen bee, introduction of a, 287
Limits of vision with ants, experiments as to, 199–206; in Daphnia, 219
Lincecum, Dr., as to Texan harvesting ants, 62
Linnæus quoted as to aphides, 67
Locust, ants apparently deceived by a leaf-like, 66
Lomechusa fed by ants, 76
Long, Col., as to the sense of hearing among certain ants, 226
Longevity of workers of *Lasius niger*, 38, 42; of queen ants, 9, 40
Lowne, Mr., quoted as to the functions of ocelli, 183
Lund, M., quoted as to the intelligence of ants, 236
Lycæna pseudargiolus, ants seen licking the larva of, 68

McCOOK, Mr., quoted as to the adoption of a queen by *Crematogaster*, 34; as to honey-ants, 48; as to the grain-fields of the Texan harvesting ant, 62; as to ants licking the larva of a butterfly, 68; on recognition by smell among ants, 127
Märkel quoted as to insects kept by *F. rufa*, 74
Maimonides as to the ownership of ant-stores of grain, 59
Males only produced by eggs laid by workers among bees, wasps, and ants, 36, 37; of *Anergates*

MYR

wingless, 86; of *Tomognathus* wingless, 87 *note*
Mandibles of ants, 11; pointed in *Polyergus*, 18; but toothless, 82; sabre-like in *S. Huberi*, 84
Marking ants, bees, and wasps. methods of, 5
Meer Hassan Ali, as to harvesting ants, 60
Mental powers of ants differ from those of men in degree rather than in kind, 181
Metamorphoses undergone by insects, 8
'Metamorphoses of Insects,' quoted, 30
Mexico, honey-pot ants in, 19, 47
Michael, Mr., description of *Uropoda* by, 429
Microphone, experiment with, 225
Mimicry, protective, instances of, 66
Mischna, rules in, respecting ant-hoards of grain, 59
Mites, ants infested by, 26, 98; new species in nests of *Lasius flavus*, 429
Mocquerys, M., on the tenacity of the bite of the ant, 96
Moggridge, Mr., on harvesting ants, 61
Morality among ants, question as to, 93; among bees, 285
Mosaic theory of the vision of compound eyes, 184
Moseley, Mr. H. N., quoted as to the connection between ants and certain epiphytes, 58
Mouth of ant described, 11
Müller, his observations on *Claviger*, 76; on the mosaic theory of vision, 184; on the colour sense in bees, 307; on blue flowers, 310
Mushrooms grown by ants, 57
Myrmecina Latreillii, the eye in, 11; their mode of defence, 16; said to be phlegmatic in disposition, 27

Myrmecocystus mexicanus, honey-holding individuals among, 19; foragers bring supplies of honey to them, 47

Myrmecodia armata, its association with ants, 58

Myrmecophilous insects, 68-78

Myrmica ruginodis, period of larval life in, 7; length of life of males of, 9; origin of a nest of, 32; observations on a wounded specimen of, 96; experiments as to recognition among, 121; as to communication among, 164, 348; illustration of terminal portion of antenna of, 227; sense-organ in tibia of, 233; their unwillingness to face a fall, 245

— *scabrinodis*, cowardly nature of, 27; *Platyarthrus* a guest of, 75

Myrmicidæ, one of the three families of ants, 1; correlation of form of knot with power of stinging in the, 13

NECTARIES, uses of, 56

Nests for ants, artificial, described, 3, 164; of ants classified and described, 23, &c.; three modes of formation of, suggested, 30; of *Lasius niger* described and illustrated, 42

Neuters among ants, production of different forms of, 22

Normann, M. de, honey ants brought from Mexico by, 47

OCELLUS, or simple eye in ants, 10; absent in the workers of some species, 11; origin of, 182

Œcodoma cephalotes, the Saüba ant, five kinds of individuals among, 21; extent of nest of, 24; leaf-cutting among, 237; intelligence of, 239

Œcophylla, sting in, 13

Organ of sense in antenna of ant, 226; in tibia of *Gryllus*, 231; in tibia of ant, 232

Ormerod, Mr., as to the sense of hearing among wasps, 221

Ova of aphis described, 71 *note*

PARASITES of ants, 26, 67, 74, 431, 433; of bees, 26

Pass-word, experiments as to, among ants, 108; supposed use of, among bees and ants, 126; experiments as to, with pupæ, 129; existence of, apparently disproved, 147

Pets, domestic animals kept as, by savages, 77

Phases of life among men and ants, analogy between, 91

Pheidole megacephala, pugnacity of small workers of, 20; experiment as to power of communication among, 180; sense-organ in tibia of, 233

— *pallidula*, the eye in, 11; communication among, 158

— *providens*, its storage of grass seeds, 60

Phora formicarum, its attacks on ants, 26, 74; described, 432

Phoridæ parasitic on ants, 67, 74; new genus and species of, 431

Pigs kept as pets by savages, 77

Plagiolepis pygmæa, the eye in, 11

Plants, relation of ants to, 50; their different modes of defence against unprofitable insects, 51; benefited by the action of insectivorous ants, 59

Plato, epigram by, quoted, 185

Platyarthrus Hoffmanseggii, a guest of the ants, 75; experiments with, 90

Platyphora Lubbockii, 67; described 431

Pogonomyrmex barbatus, stores 'ant-rice,' 61

POI

Poison ejected by *Formica rufa*, 15
Polistes, robbery among, 286; *P. gallica*, a specimen of, kept for nine months, 315
Polyergus rufescens, the eye in, 11; its mode of combat, 18; individual courage of, 27; males produced from eggs laid by workers among, 39, 45; greatly dependent on its slaves, 80, 83; slave-making expedition of, described, 81; degrading effect of slave-holding on, 89; imprisoned friends and strangers equally neglected, 105; power of communication among, 158, 180
Polygonum amphibium, glandular hairs absent from specimens growing in water, 56
Ponera contracta, the eye in, 11
Poneridæ, one of the three families of ants, 1, form of knot in, 13; stridulating apparatus in, 230
Primulaceæ, evolution of colour in, 309
Protective mimicry, 66
Pupæ of ants, 7; experimented on as to power of recognition among ants, 129
Python said to have been destroyed by the Driver ants, 64

QUEEN ants, longevity of, 9, 41; their wings, 12; several in a nest, 19; reluctance of ants to adopt a new, 32; never produced from workers' eggs, 36; seldom produced in captivity, 40; treatment of a dead, 108
Queen bees, limited nature of devotion of subjects to, 287

*R*ANUNCULACEÆ, correlation of colour with specialisation of form in, 308
Recognition of friends by ants, experiments on, 108, &c., 119;

SIE

&c.; after long separation, 123, 233, 333; means of, 125; among bees, 126; experiments as to, with pupæ, 129–147; as to sister ants brought up separately, 147; proved to be communal, not personal, 152
Relations, behaviour of ants to, 93
Retrogression of organs: of sting, 14; of wings, 15; of eyes, 75
Roads made by ants, 25
Robbery among bees, 285
Rufescent ants on a slave-making expedition, 81

ST. FARGEAU, Lepeletier de, on the origin of ants' nests, 31; on the benevolence of ants, 94; as to hearing among insects, 221
Saüba ant, see *Œcodoma cephalotes*
Sauvages, Abbé Boisier de, on the connection between ants and aphides, 68
Savage, Rev. T. S., 'On the Habits of Driver Ants,' quoted, 20, 63, 64
Scavengers, some ant-guests may serve as, 75
Scent, power among ants of tracking by, 124, 171; experiments with different kinds of, 233; importance of, to ants, 258
Schenk, *Anergates* discovered by, 86
Secretion of aphis retained till required by ants, 69; of *Claviger* and *Dinarda* as food for ants, 75, 76
Seeds of violet collected by ants, 26; stored by ants, 60; and prevented from sprouting, 61
Senses of ants, 182; organs of, 226, 232
Sentinels among bees, 288
Sex of eggs determined by treatment, 40, 41
Siebold, von, on sense-organs in

Gryllus, 231; on robbery among *Polistes*, 286
Sight, how far ants are guided by, 251, 258, 266. 270
Signals given by ants, 158
Slave-holding, structural changes induced by, 82; degradation caused by, 89
Slave-making ants, 18; expedition of, 81
Slavery among ants, origin of, 79; degrading tendency of, 82-89
Smell of *Myrmecina* possibly protective, 17; on recognition among ants by, 127; sense of, among ants, 233; the probable means of recognition among bees, 281; sense of, keen among bees, 288
Snake killed by Driver ants, 64
Soldiers among ants, 20; those of Saüba do not fight, 22; their origin, 22
Solenopsis fugax, the eye in, 11; the enemy of its hosts, 78
Solomon on the foresight of the ant, 59, 60
Sound, apparent insensibility of ants to, 222; possible existence of, beyond human auditory range, 223, 233; how produced by *Mutilla*, 229; apparent insensibility of bees to, 290
Specialisation of form in flowers correlated with colour, 308
Spectrum, experiments as to perception of, by ants, 198
Spiders, their intelligence in escaping the Ecitons, 66; mimicking an*s, 66
Spiracles of ant, position of, 12, 14
Stenamma Woodwardii, the eye in, 11; found exclusively in nests of Formica, 78
Stethoscope like organs in antenna of ant, 228
Sting in ants, possible correlation of, with form of knot, 13; probable common origin of, in ants,

bees, and wasps, 14; atrophied condition of, in *Formica*, 15; the loss of, fatal to bees, 283
Strangers, behaviour of ants towards, 104, 109, 119. 333
Stridulating apparatus in *Mutilla*, 229; in ants, 230
Strongylognathus Huberi, its mode of slave-making, 84
— *testaceus*, the eye in, 11; slaveholders in spite of their feebleness, 84; their degradation, 87, 89
Sulphate of quinine, experiment with, as to ant vision, 216
Surgical use of ant heads in Brazil, 96
Sykes, Mr., quoted as to seedcollecting ants, 60

TAME wasp, behaviour of a, 315
Tapinoma, length of period of larval life in, 7
— *erraticum*, the eye in, 11; their agility, 24; *Hetærius* in nests of, 77
Teazle, possible uses of leaf-cup in, 52
Tetramorium cæspitum, the eye in, 11; feigns death as a defence, 17; alleged greediness of, 27; enslaved by *Strongylognathus*, 84; entire dependence of *Anergates* upon, 85
Texas, harvesting ants in, 61
Thiasophila angulata in ants' nests, 77
Thorax of ant described. 12
Tibia of *Gryllus*, sense organ in, 231; of *Lasius*, 232; of *Locustidæ*, 233
Tomognathus sublævis, only workers of, known, 87 *note*
Tracks of ants illustrated, 251-257
Tuning-forks, experiments with, 222
Tunnels formed by ants, 25
Tyndall, Professor, experiment

TYP

with sensitive flames, 225; as to sense organs in antennæ, 228
Typhlopone, absence of eyes in, 11

ULTRA-RED rays, ants not sensitive to, 206
Ultra-violet rays, sensitiveness of ants to, 201–220
Uropoda formicaria described, 429

VARIETIES produced in beetles frequenting nests of various ants, 77
Verrall, G. H., Esq., description of a new genus of *Phoridæ*, 26, 431
Vespa germanica, experiment with, as to communication, 415
— *vulgaris*, experiment as to colour with, 316
Vibrations producing sensations of sound and colour, 225
Violet light, avoidance of, by ants, 189
Violets, colours of, 309; seeds of, carried into nests by *Lasius*, 26, 59
Viscidity of plants, a defence against insects, 55, 56
Vision among ants, 182; of the

WOR

ocelli, 183; of compound eyes, 184; limits of, 199, 206, 219

WALKER, Mr., honey ant sent from Australia by, 48
Wasps, occasional fertility of workers among, 36; sense of hearing among, 221; experiments with, 311, 415; more clever than bees in finding their way, 313; their courage, 314; account of a tame one, 315; their colour sense, 316; their industry, 421
Water, ants' visits prevented by, 52
Wesmael, M., describes *Myrmecocystus*, 47
Westwood, Mr., on the production of neuters, 22; on the sound produced of *Mutilla*, 229
Wings of ants, atrophy of, among the workers, 12; pulled off after flight by the queen ants, 12, 19
Winter, aphis eggs tended by ants through the, 70
Workers among ants always wingless, 12; varieties of form among, 19–22; occasional fertility of, 35; longevity of, 87, 86

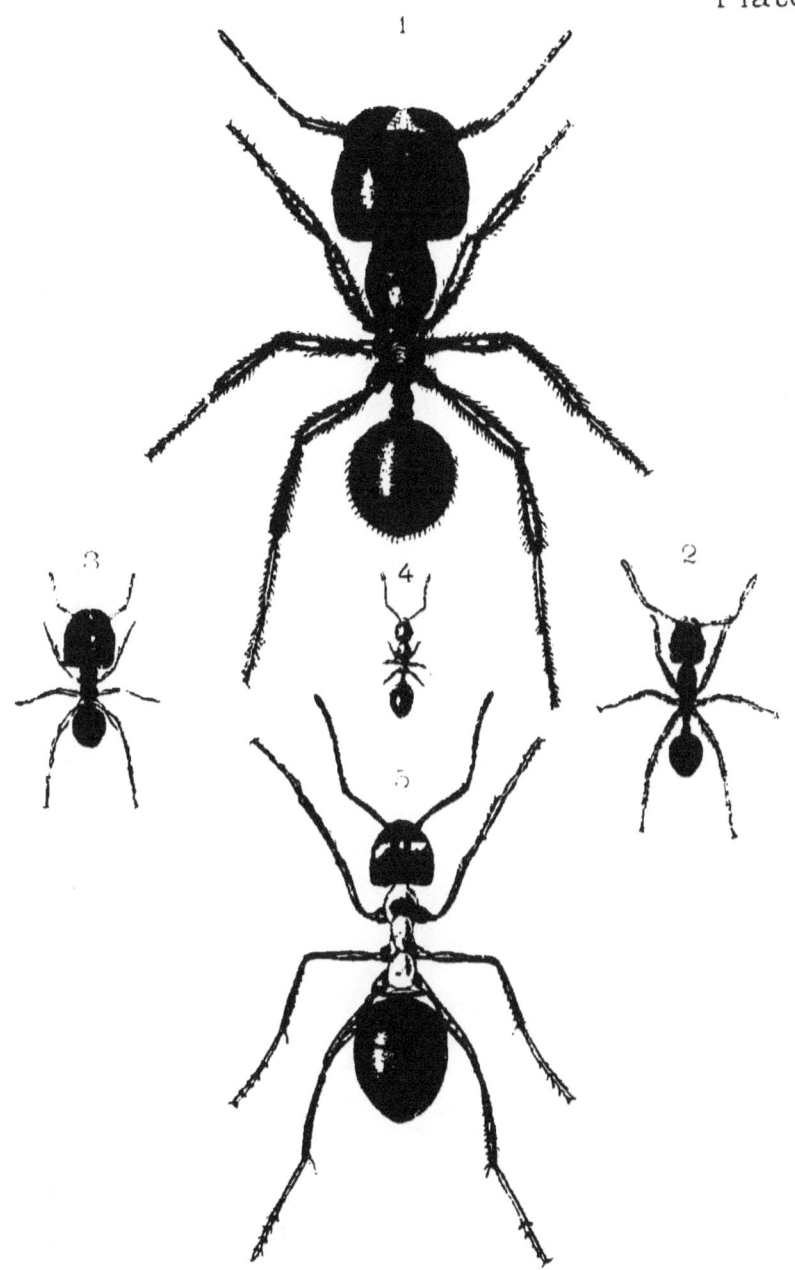

Plate 2.

1. Atta barbara ☿ major.   3. Pheidole megacephala ☿ major.
2. "   "   " minor.   4. "   "   " minor.
5. Formica rufa.

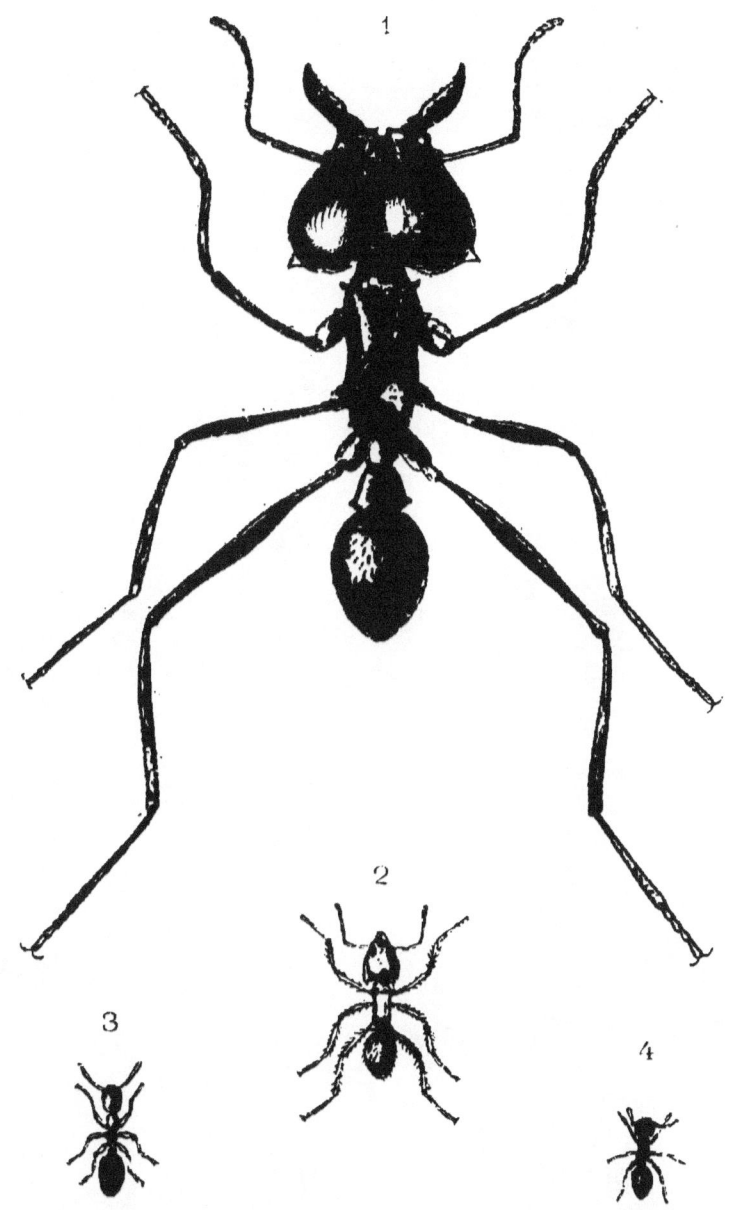

1. Œcodoma cephalotes ⚥ major.
2. "   "   " minor.
3. Stenamma Westwoodii ⚥.
4. Solenopsis fugax   "

Plate 4.

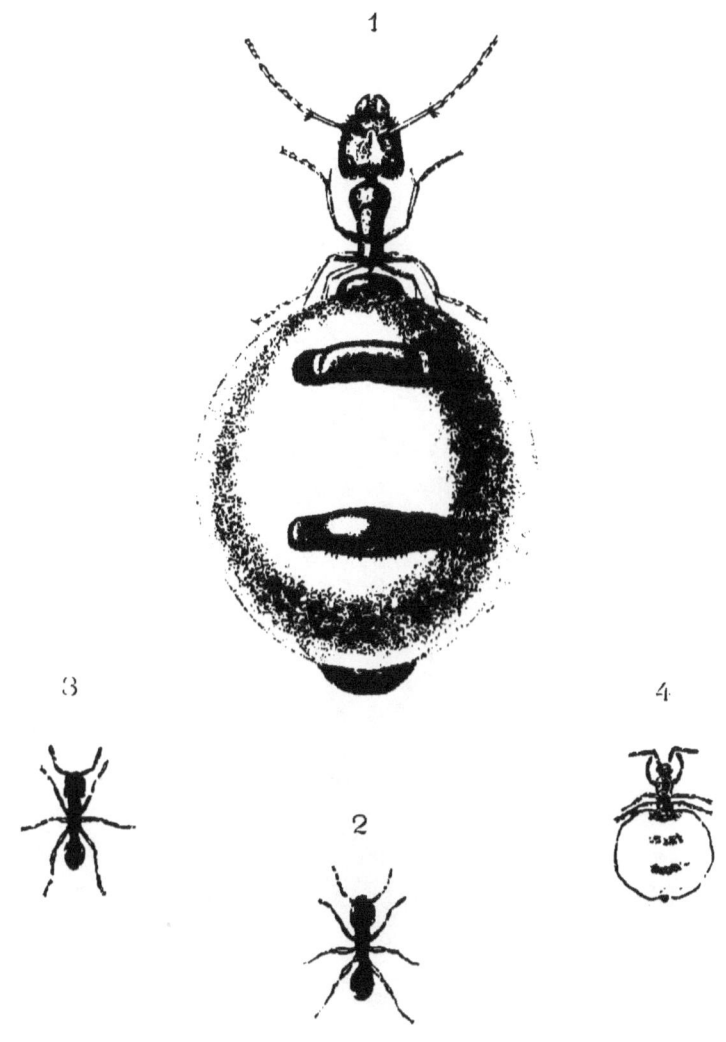

1. Camponotus inflatus ⚥.
2. Tetramorium cæspitum "
3. Strongylognathus testaceus ⚥.
4. Anergates atratulus ♀.

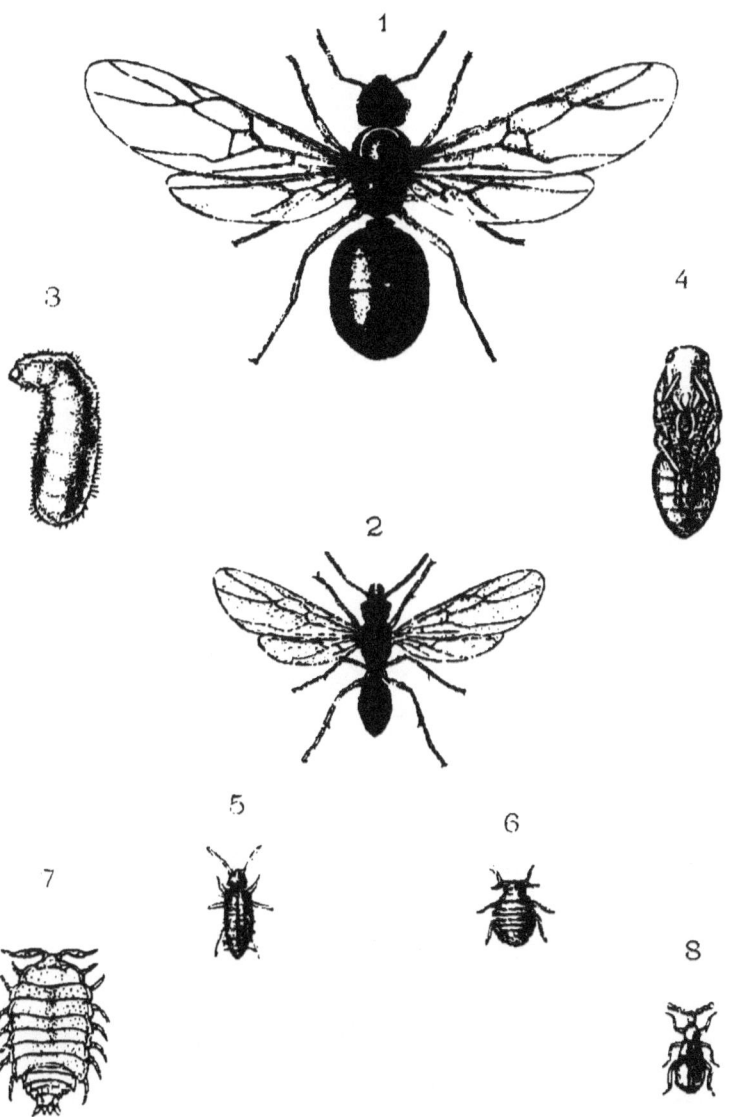

1. Lasius flavus ♀.
2. "   "   ♂.
3. "   "   Larva.
4. "   "   Pupa.
5. Beckia albinos.
6. Aphis.
7. Platyarthrus Hoffmanseggii.
8. Claviger foveolatus.

D. APPLETON AND COMPANY'S PUBLICATIONS.

Recent Volumes of the International Scientific Series.

THE AURORA BOREALIS. By Alfred Angot, Honorary Meteorologist to the Central Meteorological Office of France. With 18 Illustrations. $1.75.

While there have been many monographs in different languages upon various phases of this subject, there has been a want of a convenient and comprehensive survey of the whole field. Prof. Angot has cited a few illustrations of each class of phenomena, and without encumbering his book with a mass of minor details, he presents a picture of the actual state of present knowledge, with a summary both of definite results and of the points demanding additional investigation.

THE EVOLUTION OF THE ART OF MUSIC. By C. Hubert H. Parry, D. L. C., M. A., etc. $1.75.

Dr. Parry's high rank among modern writers upon music assures to this book a cordial welcome. It was first published as "The Art of Music," in octavo form. The title of this revised edition has been slightly amplified, with a view of suggesting the intention of the work more effectually.

WHAT IS ELECTRICITY? By John Trowbridge, S. D., Rumford Professor and Lecturer on the Applications of Science to the Useful Arts, Harvard University. Illustrated. $1.50.

Prof. Trowbridge's long experience both as an original investigator and as a teacher imparts a peculiar value to this important work. Finding that no treatise could be recommended which answers the question, What is Electricity? satisfactorily, he has explained in a popular way the electro-magnetic theory of light and heat, and the subject of periodic currents and electric waves, seeking an answer for his titular question in the study of the transformations of energy and a consideration of the hypotheses of movements in the ether.

ICE-WORK, PRESENT AND PAST. By T. G. Bonney, D. Sc., F. R. S., F. S. A., etc., Professor of Geology at University College, London. $1.50.

In his work Prof. Bonney has endeavored to give greater prominence to those facts of glacial geology on which all inferences must be founded. After setting forth the facts shown in various regions, he has given the various interpretations which have been proposed, adding his comments and criticisms. He also explains a method by which he believes we can approximate to the temperature at various places during the Glacial epoch, and the different explanations of this general refrigeration are stated and briefly discussed.

MOVEMENT. By E. J. Marey, Member of the Institute and of the Academy of Medicine; Professor at the College of France; Author of "Animal Mechanism." Translated by Eric Pritchard, M. A. With 200 Illustrations. $1.75.

The present work describes the methods employed in the extended development of photography of moving objects attained in the last few years, and shows the importance of such researches in mechanics and other departments of physics, the fine arts, physiology and zoölogy, and in regulating the walking or marching of men and the gait of horses.

D. APPLETON AND COMPANY, NEW YORK.

D. APPLETON AND COMPANY'S PUBLICATIONS.

New Volumes in the International Education Series.

FROEBEL'S EDUCATIONAL LAWS FOR ALL TEACHERS. By James L. Hughes, Inspector of Schools, Toronto. Vol. 41, International Education Series. 12mo. Cloth, $1.50.

The aim of this book is to give a simple exposition of the most important principles of Froebel's educational philosophy, and to make suggestions regarding the application of these principles to the work of the schoolroom in teaching and training. It will answer the question often propounded, How far beyond the kindergarten can Froebel's principles be successfully applied?

SCHOOL MANAGEMENT AND SCHOOL METHODS. By Dr. J. Baldwin, Professor of Pedagogy in the University of Texas; Author of "Elementary Psychology and Education" and "Psychology applied to the Art of Teaching." Vol. 40, International Education Series. 12mo. Cloth, $1.50.

This is eminently an everyday working book for teachers; practical, suggestive, inspiring. It presents clearly the best things achieved, and points the way to better things. School organization, school control, and school methods are studies anew from the standpoint of pupil betterment. The teacher is led to create the ideal school, embodying all that is best in school work, and stimulated to endeavor earnestly to realize the ideal.

PRINCIPLES AND PRACTICE OF TEACHING. By James Johonnot. Revised by Sarah Evans Johonnot. 12mo. Cloth, $1.50.

This book embodies in a compact form the results of the wide experience and careful reflection of an enthusiastic teacher and school supervisor. Mr. Johonnot as an educational reformer helped thousands of struggling teachers who had brought over the rural school methods into village school work. He made life worth living to them. His help, through the pages of this book, will aid other thousands in the same struggle to adopt the better methods that are possible in the graded school. The teacher who aspires to better his instruction will read this book with profit.

THE INTELLECTUAL AND MORAL DEVELOPMENT OF THE CHILD. Containing the Chapters on Perception, Emotion, Memory, Imagination, and Consciousness. By Gabriel Compayré. Translated from the French by Mary E. Wilson, B. L. Smith College, Member of the Graduate Seminary in Child Study, University of California. $1.50.

The object of the present work is to bring together in a systematic, pedagogical form what is known regarding the development of infant children, so far as the facts have any bearing upon early education. It contains the chapters on Perception, Emotion, Memory, Imagination, and Consciousness. Another volume will follow, completing the work, and discussing the subjects of Judgment, Learning to Talk, Activity, Moral Sense, Character, Morbid Tendencies, Selfhood, and Personality.

D. APPLETON AND COMPANY, NEW YORK.

D. APPLETON AND COMPANY'S PUBLICATIONS.

THE ANTHROPOLOGICAL SERIES.

NOW READY.

THE BEGINNINGS OF ART. By ERNST GROSSE, Professor of Philosophy in the University of Freiburg. A new volume in the Anthropological Series, edited by Professor FREDERICK STARR. Illustrated. 12mo. Cloth, $1.75.

This is an inquiry into the laws which control the life and development of art, and into the relations existing between it and certain forms of civilization. The origin of an artistic activity should be sought among the most primitive peoples, like the native Australians, the Mincopies of the Andaman Islands, the Botocudos of South America, and the Eskimos; and with these alone the author studies his subject. Their arts are regarded as a social phenomenon and a social function, and are classified as arts of rest and arts of motion. The arts of rest comprise decoration, first of the body by scarification, painting, tattooing, and dress; and then of implements—painting and sculpture; while the arts of motion are the dance (a living sculpture), poetry or song, with rhythm, and music.

WOMAN'S SHARE IN PRIMITIVE CULTURE. By OTIS TUFTON MASON, A. M., Curator of the Department of Ethnology in the United States National Museum. With numerous Illustrations. 12mo. Cloth, $1.75.

"A most interesting *résumé* of the revelations which science has made concerning the habits of human beings in primitive times, and especially as to the place, the duties, and the customs of women."—*Philadelphia Inquirer.*

THE PYGMIES. By A. DE QUATREFAGES, late Professor of Anthropology at the Museum of Natural History, Paris. With numerous Illustrations. 12mo. Cloth, $1.75.

"Probably no one was better equipped to illustrate the general subject than Quatrefages. While constantly occupied upon the anatomical and osseous phases of his subject, he was none the less well acquainted with what literature and history had to say concerning the pygmies. . . . This book ought to be in every divinity school in which man as well as God is studied, and from which missionaries go out to convert the human being of reality and not the man of rhetoric and text-books."—*Boston Literary World.*

THE BEGINNINGS OF WRITING. By W. J. HOFFMAN, M. D. With numerous Illustrations. 12mo. Cloth, $1.75.

This interesting book gives a most attractive account of the rude methods employed by primitive man for recording his deeds. The earliest writing consists of pictographs which were traced on stone, wood, bone, skins, and various paperlike substances. Dr. Hoffman shows how the several classes of symbols used in these records are to be interpreted, and traces the growth of conventional signs up to syllabaries and alphabets—the two classes of signs employed by modern peoples.

IN PREPARATION.

THE SOUTH SEA ISLANDERS. By Dr. SCHMELTZ.
THE ZUÑI. By FRANK HAMILTON CUSHING.
THE AZTECS. By Mrs. ZELIA NUTTALL.

D. APPLETON AND COMPANY, NEW YORK.

D. APPLETON & CO.'S PUBLICATIONS.

THE BEGINNERS OF A NATION. A History of the Source and Rise of the Earliest English Settlements in America, with Special Reference to the Life and Character of the People. The first volume in A History of Life in the United States. By EDWARD EGGLESTON. Small 8vo. Cloth, gilt top, uncut, with Maps, $1.50.

"Few works on the period which it covers can compare with this in point of mere literary attractiveness, and we fancy that many to whom its scholarly value will not appeal will read the volume with interest and delight."—*New York Evening Post.*

"Written with a firm grasp of the theme, inspired by ample knowledge, and made attractive by a vigorous and resonant style, the book will receive much attention. It is a great theme the author has taken up, and he grasps it with the confidence of a master."—*New York Times.*

"Mr. Eggleston's 'Beginners' is unique. No similar historical study has, to our knowledge, ever been done in the same way. Mr. Eggleston is a reliable reporter of facts; but he is also an exceedingly keen critic. He writes history without the effort to merge the critic in the historian. His sense of humor is never dormant. He renders some of the dullest passages in colonial annals actually amusing by his witty treatment of them. He finds a laugh for his readers where most of his predecessors have found yawns. And with all this he does not sacrifice the dignity of history for an instant."—*Boston Saturday Evening Gazette.*

"The delightful style, the clear flow of the narrative, the philosophical tone, and the able analysis of men and events will commend Mr. Eggleston's work to earnest students."—*Philadelphia Public Ledger.*

"The work is worthy of careful reading, not only because of the author's ability as a literary artist, but because of his conspicuous proficiency in interpreting the causes of and changes in American life and character."—*Boston Journal.*

"It is noticeable that Mr. Eggleston has followed no beaten track, but has drawn his own conclusions as to the early period, and they differ from the generally received version not a little. The book is stimulating and will prove of great value to the student of history."—*Minneapolis Journal.*

"A very interesting as well as a valuable book. . . . A distinct advance upon most that has been written, particularly of the settlement of New England."—*Newark Advertiser.*

"One of the most important books of the year. It is a work of art as well as of historical science, and its distinctive purpose is to give an insight into the real life and character of people. . . . The author's style is charming, and the history is fully as interesting as a novel."—*Brooklyn Standard-Union.*

"The value of Mr. Eggleston's work is in that it is really a history of 'life,' not merely a record of events. . . . The comprehensive purpose of his volume has been excellently performed. The book is eminently readable."—*Philadelphia Times.*

New York: D. APPLETON & CO., 72 Fifth Avenue.

D. APPLETON & CO.'S PUBLICATIONS.

THE RISE AND GROWTH OF THE ENGLISH NATION. With Special Reference to Epochs and Crises. A History of and for the People. By W. H. S. AUBREY, LL. D. In Three Volumes. 12mo. Cloth, $4.50.

"The merit of this work is intrinsic. It rests on the broad intelligence and true philosophy of the method employed, and the coherency and accuracy of the results reached. The scope of the work is marvelous. Never was there more crowded into three small volumes. But the saving of space is not by the sacrifice of substance or of style. The broadest view of the facts and forces embraced by the subject is exhibited with a clearness of arrangement and a definiteness of application that render it perceptible to the simplest apprehension."—*New York Mail and Express.*

"A useful and thorough piece of work. One of the best treatises which the general reader can use."—*London Daily Chronicle.*

"Conceived in a popular spirit, yet with strict regard to the modern standards. The title is fully borne out. No want of color in the descriptions."—*London Daily News.*

"The plan laid down results in an admirable English history."—*London Morning Post.*

"Dr. Aubrey has supplied a want. His method is undoubtedly the right one."—*Pall Mall Gazette.*

"It is a distinct step forward in history writing; as far ahead of Green as he was of Macaulay, though on a different line. Green gives the picture of England at different times—Aubrey goes deeper, showing the causes which led to the changes."—*New York World.*

"A work that will commend itself to the student of history, and as a comprehensive and convenient reference book."—*The Argonaut.*

"Contains much that the ordinary reader can with difficulty find elsewhere unless he has access to a library of special works."—*Chicago Dial.*

"Up to date in its narration of fact, and in its elucidation of those great principles that underlie all vital and worthy history.... The painstaking division, along with the admirably complete index, will make it easy work for any student to get definite views of any era, or any particular feature of it.... The work strikes one as being more comprehensive than many that cover far more space."—*The Christian Intelligencer.*

"One of the most elaborate and noteworthy of recent contributions to historical literature."—*New Haven Register.*

"As a popular history it possesses great merits, and in many particulars is excelled by none. It is full, careful as to dates, maintains a generally praiseworthy impartiality, and it is interesting to read."—*Buffalo Express.*

"These volumes are a surprise and in their way a marvel.... They constitute an almost encyclopædia of English history, condensing in a marvelous manner the facts and principles developed in the history of the English nation.... The work is one of unsurpassed value to the historical student or even the general reader, and when more widely known will no doubt be appreciated as one of the remarkable contributions to English history published in the century."—*Chicago Universalist.*

"In every page Dr. Aubrey writes with the far reaching relation of contemporary incidents to the whole subject. The amount of matter these three volumes contain is marvelous. The style in which they are written is more than satisfactory.... The work is one of unusual importance."—*Hartford Post.*

New York: D. APPLETON & CO., 72 Fifth Avenue.

D. APPLETON & CO.'S PUBLICATIONS.

JOHN BACH MCMASTER.

HISTORY OF THE PEOPLE OF THE UNITED STATES,

from the Revolution to the Civil War. By JOHN BACH MCMASTER. To be completed in six volumes. Vols. I, II, III, and IV now ready. 8vo. Cloth, gilt top, $2.50 each.

". . . Prof. McMaster has told us what no other historians have told. . . . The skill, the animation, the brightness, the force, and the charm with which he arrays the facts before us are such that we can hardly conceive of more interesting reading for an American citizen who cares to know the nature of those causes which have made not only him but his environment and the opportunities life has given him what they are."—*N. Y. Times.*

"Those who can read between the lines may discover in these pages constant evidences of care and skill and faithful labor, of which the old-time superficial essayists, compiling library notes on dates and striking events, had no conception; but to the general reader the fluent narrative gives no hint of the conscientious labors, far-reaching, world-wide, vast and yet microscopically minute, that give the strength and value which are felt rather than seen. This is due to the art of presentation. The author's position as a scientific workman we may accept on the abundant testimony of the experts who know the solid worth of his work; his skill as a literary artist we can all appreciate, the charm of his style being self-evident."—*Philadelphia Telegraph.*

"The third volume contains the brilliantly written and fascinating story of the progress and doings of the people of this country from the era of the Louisiana purchase to the opening scenes of the second war with Great Britain—say a period of ten years. In every page of the book the reader finds that fascinating flow of narrative, that clear and lucid style, and that penetrating power of thought and judgment which distinguished the previous volumes."—*Columbus State Journal.*

"Prof. McMaster has more than fulfilled the promises made in his first volumes, and his work is constantly growing better and more valuable as he brings it nearer to our own time. His style is clear, simple, and idiomatic, and there is just enough of the critical spirit in the narrative to guide the reader."—*Boston Herald.*

"Take it all in all, the History promises to be the ideal American history. Not so much given to dates and battles and great events as in the fact that it is like a great panorama of the people, revealing their inner life and action. It contains, with all its sober facts, the spice of personalities and incidents, which relieves every page from dullness."—*Chicago Inter-Ocean.*

"History written in this picturesque style will tempt the most heedless to read. Prof. McMaster is more than a stylist; he is a student, and his History abounds in evidences of research in quarters not before discovered by the historian."—*Chicago Tribune.*

"A History *sui generis* which has made and will keep its own place in our literature."—*New York Evening Post.*

"His style is vigorous and his treatment candid and impartial."—*New York Tribune.*

New York : D. APPLETON & CO., 72 Fifth Avenue.

D. APPLETON & CO.'S PUBLICATIONS.

THE INTELLECTUAL RISE IN ELECTRICITY. A History. By PARK BENJAMIN, Ph.D., LL.B., Member of the American Institute of Mechanical Engineers, Associate Member of the Society of Naval Architects and Marine Engineers, etc. With Three Portraits. 8vo. Cloth, $4.00.

"Mr. Benjamin surely has produced a book that will find interested readers throughout the entire world, for wherever electricity goes as a commercial commodity a desire to know of its discovery and development will be awakened, and the desire can be satisfied through no more delightful channel than through the information contained in this book."—*New York Times.*

"Mr. Benjamin has performed his self imposed task in an admirable fashion, and has produced a work which has a distinct historical value."—*Brooklyn Eagle.*

"A work that takes a high rank as a history dealing with an abstruse topic, but bestowing on it a wealth of vital interest, pouring over it streams of needed light, and touching all with a graceful literary skill that leaves nothing to be desired."—*New York Mail and Express.*

"A very comprehensive and thorough study of electricity in its infancy. He presents his matter clearly and in an interesting form. His volume is one of especial value to the electrical student, and the average reader will read it with interest."—*Milwaukee Journal.*

"The work is distinctly a history. No technical preparation is required to read it, and it is free from all mathematical or other discussions which might involve difficulty. The style is, in the main, excellent."—*Science.*

"A remarkable book. . . . A book which every electrician ought to have at hand for reference—historic, not scientific reference—and which will prove instructive reading to the thoughtful of all classes."—*New York Herald.*

"The most complete and satisfactory survey of the subject yet presented to the reading public. . . . A volume which will appeal to an ever-increasing body of people; and as a reference book it will prove invaluable to writers on the development and utility of electricity."—*Philadelphia Evening Bulletin.*

"The leading work on the subject in any language."—*New York Evening Post.*

"One of the best works devoted to the development of the great force of modern time that has been published in the last decade."—*New York Commercial Advertiser.*

"The author has written a plain and simple history of the beginnings of electrical science, none the less but rather the more valuable because, without dilution or sacrifice of accuracy, he has excluded mere technicalities and gratuitous scientific demonstrations."—*Philadelphia Press.*

New York: D. APPLETON & CO., 72 Fifth Avenue.

D. APPLETON & CO.'S PUBLICATIONS.

THE LIBRARY OF USEFUL STORIES.

Each book complete in itself. By writers of authority in their various spheres. 16mo. Cloth, 40 cents per volume.

NOW READY.

THE STORY OF THE STARS. By G. F. CHAMBERS, F. R. A. S., author of "Handbook of Descriptive and Practical Astronomy," etc. With 24 Illustrations.

"The author presents his wonderful and at times bewildering facts in a bright and cheery spirit that makes the book doubly attractive."—*Boston Home Journal.*

THE STORY OF "PRIMITIVE" MAN. By EDWARD CLODD, author of "The Story of Creation," etc.

"No candid person will deny that Mr. Clodd has come as near as any one at this time is likely to come to an authentic exposition of all the information hitherto gained regarding the earlier stages in the evolution of mankind."—*New York Sun.*

THE STORY OF THE PLANTS. By GRANT ALLEN, author of "Flowers and their Pedigrees," etc.

"As fascinating in style as a first-class story of fiction, and is a simple and clear exposition of plant life."—*Boston Home Journal.*

THE STORY OF THE EARTH. By H. G. SEELEY, F. R. S., Professor of Geography in King's College, London. With Illustrations.

"It is doubtful if the fascinating story of the planet on which we live has been previously told so clearly and at the same time so comprehensively."—*Boston Advertiser.*

THE STORY OF THE SOLAR SYSTEM. By G. F. CHAMBERS, F. R. A. S.

"Any intelligent reader can get clear ideas of the movements of the worlds about us. ... Will impart a wise knowledge of astronomical wonders."—*Chicago Inter-Ocean.*

THE STORY OF A PIECE OF COAL. By E. A. MARTIN, F. G. S.

"The value and importance of this volume are out of all proportion to its size and outward appearance."—*Chicago Record.*

THE STORY OF ELECTRICITY. By JOHN MUNRO, C. E.

"The book is an excellent one, crammed full of facts, and deserves a place not alone on the desk of the student, but on the workbench of the practical electrician."—*New York Times.*

THE STORY OF EXTINCT CIVILIZATIONS OF THE EAST. By ROBERT ANDERSON, M. A., F. A. S., author of "Early England," "The Stuart Period," etc.

New York: D. APPLETON & CO., 72 Fifth Avenue.

www.ingramcontent.com/pod-product-compliance
Lightning Source LLC
Chambersburg PA
CBHW051844300426
44117CB00006B/267